Gilgamesh in the 21st Century

A Personal Quest to Understand Mortality

PAUL BRACKEN

CONTENTS

⌘

Preface
vii

Gilgamesh
1

An Unpalatable Fate
29

A Leap of Faith
53

Being Human
83

Synthetic Dreams
105

Wishful Thinking
137

God's Country
155

The Age of Death
193

Bibliography
235

Index
251

Notes
261

ABOUT THE AUTHOR

For more than eleven years, Paul Bracken served as Ireland's Regional Coordinator for The Planetary Society, and worked with Carl Sagan to promote space exploration and the search for extraterrestrial intelligence. He lives in California with his wife and two children.

PREFACE

The idea for this book began to crystallize in my head sometime around the turn of the century during a lecture that I was presenting for The Planetary Society. In those days, my passion for science was directed toward getting people interested in the robotic exploration of the solar system and the search for life elsewhere, but I was intrigued by how frequently people would ask about God, and the meaning of life. There were usually a few pertinent queries on the presentation itself – maybe something about the cratering patterns on Saturn's moons, or the likelihood of life on Europa – but invariably God was ushered into the proceedings. People wanted to know if science and religion were compatible. One gentleman inquired as to whether the complexity of the Sun was proof that God existed. Others asked me to speculate on what our fate may be. Why would anyone expect an astronomy presenter to be well versed in such matters?

It could be because astronomers are used to dealing with distances and timescales that are somewhat intimidating to the rest of us. A hundred million years is a relatively short time in the context of planetary evolution but when compared to the span of a human life, it seems like an eternity. As mortal beings, the cosmic perspective makes it difficult for us to feel worthwhile. Carl Sagan's advice was to "Do something worthwhile," which was a clever if not altogether satisfying response.

Generally, I was discouraged from talking about gods and the like since I was supposed to be representing The Planetary Society and there were no supernatural phenomena mentioned in our mission statement. Now that I have a few more gray hairs, I decided to review all of the material that I had collected during those years of lecturing, and it occurred to me that I had the makings of a book – one that would explore those aspects of science that people are most curious about and which have direct relevance to human existence.

I was reminded of my childhood conversations with my grandfather who had a passion for science, but who was also a deeply religious man. It seemed that while science was useful for everyday matters, questions such as "Why must I die?" or "How can I feel worthwhile?" could only be addressed by religion. As a physicist later told me, "Some things remain outside of science."

For me though, the religious approach never really worked. When I was eleven years old, I decided that I wanted to figure things out for real. I imagine I was about as determined as the ancient king Gilgamesh when he set out on his quest with the same set of concerns on his mind. "Must I die?" asked Gilgamesh. Forty five centuries later, we're still asking the same questions.

Wouldn't it be fun, I thought, to do a twenty first century reboot of Gilgamesh's quest? What might he find if he could begin his quest anew, armed with the tools of science and centuries of acquired knowledge? What could be a more interesting theme than the whole matter of life and death itself? And so, in this book, I embark on a personal voyage of discovery to find the answers to my childhood questions about the human condition, and what it means to be mortal. I invite the reader to think about death, not out of any desire to be morbid, but rather because it opens the door to a lot of interesting science, and because our mortality is often what prompts us to contemplate the grander mysteries of life. If, like my dad, you're tempted to jump directly to last paragraph, remember that it's the journey that counts – not the ending.

1

⌘

Gilgamesh

In the 1980s, I was happy to learn that I lived less than an hour's drive away from one of the most impressive Neolithic constructions in existence, a prehistoric dome shaped mound known as Newgrange. It is thought to have been built around 3,200 BCE, making it more than a thousand years older than Stonehenge and older in fact, than the more famous pyramids of Egypt.

Unlike other Irish monuments, which often amount to nothing more than a mere jumble of rocks, Newgrange is an actual building. There's just one entrance, beyond which is a narrow, nineteen meter long corridor that leads to a cruciform chamber, large enough to accommodate about twenty people, and embedded within 200,000 tonnes of rock. This inner chamber remains in darkness throughout the year, except at sunrise on the shortest days of winter.

Then, for a brief period, the darkness is disturbed by a finger of golden light which creeps along the floor of the passage, bringing a warm glow to the cave-like interior. The light enters through a small window just above the entrance, by what appears to be a carefully planned alignment with the southernmost position of the rising Sun. Thus the event marks the beginning of a new year, as the Sun heads slowly northward again.

Being naturally interested in things astronomical, my friends and I were intrigued by accounts of this solar alignment, and so on the eve of December 21st, we packed our cameras and set off into the night, arriving in the frost covered Boyne valley at about two hours past midnight. In retrospect, I will agree that three so-called astronomy experts should have known better than to anticipate a sunrise at that hour of the morning. It was a miscalculation that made for a somewhat uncomfortable night as we sat for several hours in the cold next to our five thousand year old megalithic mound, keeping a hopeful eye on the horizon. When daybreak came, we watched as the more enlightened members of the community stepped out of their cars, nice and refreshed after a good night's sleep. The Sun, they told us, was expected to peek over the hill sometime around 9:00 a.m.

There are other examples of ancient monuments with solar alignments, such as those at Chaco Canyon, New Mexico, but the Newgrange site is arguably the most enthralling, and not just because of its age. Those fortunate enough to have witnessed the event describe the experience as moving and uplifting. The architects evidently had a strong interest in the winter solstice, as a practical matter perhaps, to highlight the start of a new year, or maybe for some celebratory purpose. Five thousand years later, the solstice is still a time for celebration throughout the world. In fact, it is thought that the early Christian church invented Christmas in an attempt to supplant the pagan religions of the time, by marking the birth of Christ on a date that coincided with the ancient solstice festivities.

Our guide told us that Newgrange was built primarily for use as a tomb; the main chamber has three side recesses housing large stone basins where the cremated remains of the dead are believed to have been placed. We learned that some of the stone material used in the construction was retrieved from locations up to forty miles away, and that the entire project may have spanned several generations. What, I wondered, could have been the motivation for such a huge expenditure of effort and resources? Why build such an overly elaborate structure merely to mark the calendar, or as a place to store the ashes of the deceased?

Whatever their particular motivation might have been, it is clear that a significant amount of determination and purpose was involved, and sustained over a long period of time. Traditionally, such purpose is derived from religious belief. If this truly was a tomb, then the arrival of the solstice light might have been seen in a context related to their religious interpretations of death, just as the architecture of the pyramids was governed by the Egyptian belief in an afterlife.

Nature provides many metaphors of resurrection, such as the

of Newgrange as well, though they didn't tell us this directly.

Even today, these central questions concerning our existence have not lost their relevance. We are still bound by the rules that apply to mortal beings, we face the same destiny as Gilgamesh, and we often react as he did. For example, in the Sumerian poem "Gilgamesh and Huwawa", Gilgamesh decides that he will deal with his fear of death by attempting to leave behind a great legacy, something which is also common in modern philanthropy.[6] Consider the words of actor Leonard Nimoy, as he describes what death means on a personal level:

> I think about the loss of relationships, the end of that. I think about the loss of creative opportunity, which I love – to be creative, to see things evolve that you dream about and can bring to fruition. I think it is important now to be making philanthropic statements, to be giving back ... as much as I can, as much as we can into the community, into funding for arts, funding for children, funding for education, things that we believe in, care about. And leaving a legacy, I'm concerned about that. I'm concerned that it be positive and constructive and meaningful.[7]

So we have good evidence that from the beginning of history, death and immortality have plagued us as human beings. We are moved by Gilgamesh's fears; we can empathize with him as he describes his fear of death, as he demands to know how he can acquire immortality. Unfortunately, this precious goal remains beyond his reach. He feels the terror of death, searches desperately for a way to avoid the oblivion it promises, until finally he is forced to come to terms with his humanity, learning to make the most of life for what it is.

Instead of immortality, he obtains a deeper understanding of what it means to be a human being. The road to self-discovery is as difficult for us to traverse as it was for Gilgamesh, and it sometimes requires the same kind of heroic resolve. Despite the many challenges overcome by civilization, the world still struggles to come to grips with the problem of death. In 2005, for example, the question of whether to continue to provide life support for a brain damaged woman captured the attention of the international news media. Members of congress debated the issue, the

Vatican had something to say, and a US president was roused from his sleep to deal with the matter.[8]

There is nothing esoteric about dying. In time, we are all given a chance to know the feelings of sadness, anger or hopelessness that the total loss of an acquaintance or loved one can bring, or the trepidation that comes from contemplating our own mortality. Although we are universally afflicted by this torment, the intense grief of the bereaved is always deeply personal, and difficult to share with others, a situation which can be exacerbated by western society's façade of ambivalence toward death.

I was about twelve years old when the experience of receiving a general anesthetic allowed me to grasp the concept of unconsciousness, and what it might mean to die. I was propped up on my elbows, watching the anesthetist apply the needle to my arm, and then in what seemed like the blink of an eye, I was lying in bed, recovering from the operation. This was considerably more dramatic than the normal experience of drifting off to sleep, where the boundary between wakefulness and unconsciousness is somewhat fuzzy, and not as noticeable. It was for me, a powerful demonstration of the fragility of existence – one minute you are there, and the next, you are not.

As a child, I pondered the significance of aging and its inevitable consequences. I could think of nothing more serious in life than the intrusion of death. From the world of adults I learned that there was no greater concern than the loss of human life. Catastrophes were measured by the numbers of lives forfeited. My first close up view of the effects of death followed the loss of my mother's father, Granddad Waine. My dad came to fetch me from school and brought me over to my grandmother's house and I still have a vivid memory of her as she sat on the sofa, handkerchief in hand, along with several of my aunts, all of them in tears. I will never forget the distraught bewilderment in my granny's eyes and how she seemed to stare right through me, as though she had lost her mind. It was clear that the adults did not really have a handle on death.

Given the trauma that death brings, and how inappropriate a fate it seems for a sentient mind, the fact that humans can do nothing about it has made our story a tragic one. Tragedy is limp without death, a fact which has not been lost on the entertainment business. Its impact and importance are clearly recognized in our religions and in our storytelling, such that it plays a dominant role in human culture. Even when modern mythology presents outlandish ideas about alien life and future technologies, the stories deal primarily with the tragedy of death and mortality.

In the popular *Star Trek* show, for example, although the stories take place in the future, the space travelers are still quite mortal – death still prevails. I find it interesting that in many of these stories, although there are opportunities to escape death, the heroes often choose not to extend their natural lives. Perhaps in order to appeal to an audience of mortals, they are forced to make a case for mortality. Indeed it would be challenging for script writers to maintain the necessary level of tension if death were written out of the storyline.

But setting fantasy aside, how do we really feel about death? Is it something that has a place in society, to provide meaning and context for life? Or is the goal to banish death, through our ongoing crusade against its causes? Is it our hope that when we finally have a complete understanding of the biochemistry of life, along with an ability to engineer biological materials and circumvent disease, that we will prolong human life indefinitely? Or perhaps, through an understanding of the consciousness mind, might we find other ways to create or preserve sentient beings? Outside of fiction and futurism, the reality of dying seems to be a subject that people prefer to avoid. The whole area is often considered distasteful, perhaps because it is a problem that we are not quite sure how to deal with.

Of course the story of death is older than that of humans. Long before our arrival on the planet, death had already established itself as a powerful force in the developing biosphere, playing an essential role in the provision and dispersal of energy and raw materials for the growth of organisms and forming the basis for natural selection,[ii] without which life could not have evolved. Life feeds on death.

There is death within us – that is to say from the moment of birth, a

---

[ii] Without death, evolution would not have occurred, although it is important to note that selection depends not upon the survival of individuals, but rather upon their ability to reproduce.

continuous cycle of death and replenishment in the cellular communities that compose us. Notwithstanding the replenishment of cells, the vital machinery of the human body is subject to damage from the chemistry of life itself, and the resulting slow degradation is grudgingly accepted as the cost of growing old. I expect none of us is especially concerned about the death throes of cells, but it is the final phase, the demise of whole organisms, that we often find objectionable. We know, from the *Gilgamesh* epic, that ancient minds understood the implications for human beings.

> Though no one has seen death's face or heard
> death's voice, suddenly, savagely, death
> destroys us, all of us, old or young.
> And yet we build houses, make contracts, brothers
> divide their inheritance, conflicts occur—
> as though this human life lasted forever.
> The river rises, flows over its banks
> and carries us all away, like mayflies
> floating downstream: they stare at the Sun,
> then all at once there is nothing.[9]

This rather grim view of death was surely as unpalatable to many throughout the ages as it was to Gilgamesh himself, perhaps explaining the preference in human cultures for alternative hypotheses, such as the Egyptian concept of an afterlife. It is a strong and persistent human tradition to think of the dead as though they are still living. In the 1970s, in order to better understand how people felt about death, French author Philippe Aries took a comprehensive look at how western attitudes had evolved over the previous millennium.[10] He showed how the perception of death was altered by societal changes, such as the advent of modern medicine, for example.

Although Aries was looking primarily at how our feelings about mortality had changed, ironically his work is also a good illustration of how the same fundamental mindset has persisted through time, at least insofar as attitudes toward death have been and continue to be expressed in religious or supernatural terms. It seems that any discussion or rationalization of death by our ancestors was framed exclusively within the context of some overarching religious belief, a practice which may have persisted from the depths of pre-history. If indeed this deep-rooted cultural connection between death and religion extends all the way back to our very origin, then might this way of thinking about death be derived from something very fundamental about the nature of human beings, or perhaps about what it means to be sentient?

With respect to the multiplicity of non-sentient beings inhabiting the planet, the traditional view, at least in the western world, has been that humans are superior, elitist caretakers with divine sanction, as per Genesis, to assume "dominion" over the animals.[11] In seventeenth century England, this lofty perspective on humanity was still in force, and was most eloquently expressed in Shakespeare's *Hamlet*, in which man is described as the "paragon of animals."[12] It was believed that humans possessed a number of distinguishing characteristics that were not shared with other animals, such as a sense of morality and decency, or the ability to manufacture and make use of tools. The rest of the animal kingdom was populated with amoral, soulless creatures, incapable of reason and devoid of purpose.

With the nineteenth century, however, came the discovery of natural selection and a realization of the common ancestry of all living things. It became more difficult to draw strict lines of separation between human and animal as we began to better understand our relationship with the natural world. Our primate cousins in particular have since been shown to possess many characteristics that were previously ascribed to humans alone, including a proclivity for tool-making and social behavior patterns that betray an underlying complex of motivation and emotion.

Experiments conducted in the 1980s, for example, illustrated the remarkable cognitive ability of a bonobo named Kanzi, who learned to understand spoken English and could communicate symbolically using a lexigram keyboard.[13] In 2003, a research team at the Yerkes National Primate Research Center in Atlanta showed that capuchin monkeys seem to possess a sense of fairness; they can recognize injustice and, like us, can express their indignation.[14] Scientists have even discovered the existence of chimpanzee culture by examining the traditions of geographically dispersed communities in the wild.[15]

These similarities between humans and apes are not so surprising given that we share much of our biological infrastructure and evolutionary history. Our differences it seems are not in kind, but in degree. Yet when it comes to comprehension, and awareness, and the extraordinary extent to which human culture has reshaped the planet, it is clear that we are a breed apart. We inhabit a world of our own creation, a world that is perceived through the lens of human history and achievement, a world of human interaction that is described and understood entirely in human terms. In the words of anthropologist Terrence Deacon, ours is a virtual world of abstractions which for other animals remains "out of reach: inconceivable."[16]

We do not know precisely when our ancestors acquired the mental faculties that distinguished them from the rest of nature, but it certainly had occurred by the time the Cro-Magnon people arrived in Europe some forty thousand years ago, as evidenced by the multiplicity of artifacts and impressive cave paintings they bequeathed to us. Hundreds of decorated caves have been discovered throughout Western Europe, with human engravings and paintings dating from about thirty thousand to ten thousand years in the past, a period which coincides with the last great ice age.

While discoveries of older artifacts hint at more ancient beginnings for symbolic art, there is no question that the Cro-Magnon people distinguished themselves with a particular exuberance for symbolism and self-expression, a propensity that provides strong evidence for an underlying substrate of mental patterns and reasoning.[17]

The artistry and accuracy of these ice-age painters is noteworthy. Paleontologist Ian Tattersall admits that only through Cro-Magnon art do we know, "for example, that the extinct rhinoceroses of Ice Age Europe were adorned with shaggy coats, and that the extraordinary *Megaloceros giganteus*, a deer with vast antlers whose most recent bones date from 10,600 years ago, bore a dramatic and darkly colored hump behind the shoulders."[18] "The frozen carcasses of extinct woolly mammoths found in the wastes of Siberia, serves also to emphasize," Tattersall says, "the perceptiveness of Ice Age artists," whose cave drawings capture the physical characteristics of these creatures "right down to their remarkable split-tipped trunks."[19]

Some Cro-Magnon creations appear to be more than just faithful accounts of reality. As human observers ourselves, we can detect hidden qualities, intuitive observations, or feelings perhaps, like the perception of affection or beauty, things that only existed within the mind of the artist. One famous example at the Font de Gaume cave in Dordogne, France, depicts two reindeer with their heads bowed, facing each other, the male reindeer gently caressing the female's forehead with his tongue in what appears to be a moment of tenderness between them.

Clearly this painting represents more than just a bland account of everyday matters, such as hunting or searching for food. Even the earliest Cro-Magnon people, the Aurignacian, produced artifacts that bear witness to a culture rich in mythology, imagination and symbolism dating back well beyond thirty thousand years ago.

The "Lion Man", a thirty thousand year old carving of a human with the head of a lion, found at the Hohlenstein-Stadel site in Germany, is obviously some kind of mythological or imaginative figure, since it bears no resemblance to any creature found in nature. Interestingly though, when my daughter, who was three years old at the time, came across a photograph of this artifact, she immediately identified it as a "lion person" and demanded to know where it had come from – a thought-provoking example of communication across the millennia.

The Aurignacian culture also produced musical instruments, such as bone flutes, as well as a plentiful supply of decorative beads and bone fragments carved into animal shapes, or engraved with curious notations. Personal ornamentation was the order of the day, as necklaces made from perforated bits of stone, bones, seashells, and pierced animal teeth were common. There is even evidence to suggest that tattooing was practiced.[20]

Since much of this art was as useless in practical terms as it appears remarkable, why did people preoccupy themselves with it? In *Secrets of the Ice Age*, Evan Hadingham describes how magical rituals associated with the remote cave paintings of Australian Aborigines suggested to archaeologists that our ancient ancestors might have been similarly motivated by magic.[21] Cave art was often carried out in dark, almost inaccessible places, deep underground, where artists used stone lamps to provide artificial light, and scaffolding to reach greater heights on the cave walls and ceilings.[22]

There is little evidence that people actually lived in these deep recesses, leading to the assumption that these were sacred shrines, or mystical places, visited only for special reasons related to their art. "At one cave in France," says Tattersall, "you have to paddle up an underground stream, then walk, wriggle, and crawl for two hours before finally, a mile underground, reaching the terminal chamber."[23]

As a result of these findings, most prehistorians are now convinced that Paleolithic art represented more than just idle doodling, and strongly suspect that magical motives were behind these activities. Consequently there has been much speculation about the possible mythologies, religions, and reasoning underlying Cro-Magnon art – an exercise which Hadingham admits "has not advanced far".[24]

In the absence of the artists themselves, much of the original meaning may remain elusive. This is certainly the case where more abstract symbols are used as with any kind of written language, whose message may be entirely lost along with its creators. Hadingham notes that "It is intriguing" that abstract symbols are found even in the earliest Aurignacian art.[25] Usually these are in the form of various dots, shapes and lines that seem to be associated with animal representations.

Enigmatic signs can be found in the seventeen thousand year old paintings of the famous Lascaux caves in France, while bone plaques from the nearby site of Abri Blanchard are inscribed with symbolic notation that suggests to some researchers the possibility that they were used as notebooks, or even lunar calendars.[26] The symbolism in this ancient art reflects a mental representation of the world, a way of modeling reality that allowed our ancestors to relate to their surroundings in new and interesting ways, provided them with insight into the workings of nature, and facilitated reasoning.

For such cognitive abilities to have persisted in humans, they must have provided our ancestors with some competitive advantage. All things being equal, we could argue, it is better to be smart than dumb; although for a more nuanced appreciation of how our ancestors would have profited from their cognitive supremacy, Tattersall points to the example of the San Bushmen, who, he recalls, "can tell from a bent twig or a stone overturned in a streambed what animal had passed, in which direction, how long ago, how fast it was traveling, and whether it was wounded."[27]

This is surely an effective illustration of the practical advantages associated with symbolization and language. With these abilities, our ancestors could understand and ascribe meaning to the things they observed, and from the evidence they left behind, we can tell that this is exactly what they did.

But the real advantage of having more cognitive power was that it allowed individuals to co-exist and collaborate with one another in larger groups and, as naturalist David Attenborough explains, it is for this reason that our capable brains evolved in the first place – to give us the means to deal with social complexity. Once individuals start living in groups, life gets complicated. For baboons, says Attenborough, "Even deciding where to sit is a political decision." Monkeys living in larger groups need to work a lot harder on their interpersonal skills, so they tend to have larger brains.

> In fact, the relationship [between brains and group size] is so close, that were you to give a skull to a researcher who worked with monkeys, even though they didn't know what kind of monkey it belonged to, they would be able to accurately predict the size of group in which it lived.[28]

In addition to their flint knives, arrow heads, and sewing implements, there is no question that our species developed a penchant for decoration, far beyond what would have been required for practical use. Although, as we have discussed, prehistorians are now comfortable that magic and ritual were the prime motivators for this extra-curricular activity, this still leaves us with the question, why magic? Arguably a waste of time in their own right, how did magic and superstition come to play such an important role?

According to sociologists Rodney Stark and William Sims Bainbridge, prehistoric societies would have used magic and religion to come up with rational explanations for how the world worked. In the absence of formalized religion, a belief in magic would have helped our ancestors to make predictions and account for specific outcomes. Religion, on the other hand, provides more elaborate, generalized explanations that cover a much broader perspective, but in the case of either magic or religion we are looking at a way of thinking that betrays a sense of the spiritual.

The most compelling glimpse of our ancestors' spiritual existence comes from what we can tell about their burial practices. The Sungir site, near Vladimir, Russia, presents a dramatic illustration of a Paleolithic burial, some 28,000 years ago. The remains of three individuals were discovered, two adolescents and a sixty year old male; each had been interred in an elaborate burial garment constructed of thousands of painstakingly prepared ivory beads, along with a variety of other jewelry. Researchers figure that it took about an hour to make a single bead.

They were also accompanied by an impressive assortment of artifacts and weaponry, including spears of straightened mammoth ivory. How they managed to flatten the ivory is unknown. From the symbolic expression associated with these burials, we can tell that death was greeted with a spiritual awareness; after all, there are easier ways to dispose of bodies than to go to such great lengths. In particular, says Tattersall, the Cro-Magnon ritual of burying grave goods along with the dead is recognized by anthropologists as indicating "a belief in an afterlife." Why else would they have disposed of valuable items in this way?

> It is here that we have the most ancient incontrovertible evidence for the existence of religious experience.[29]

So it would appear that, from the very first signs of sentience on the planet, religion played an important role in the lives of our ancestors and, to answer the question we posed earlier, death and religion have always gone hand in hand. The extraordinary symbolic and artistic expression by Cro-Magnon people, in diverse environments across tens of thousands of years, can best be understood as part of a system of rituals that they would have considered essential for their prosperity and survival.

The root of all of this artistic expression and cognitive power is what Deacon regards as "the most incredible gift any species ever received at the hands of evolution" - the human capacity for symbolic representation, meaning the ability to hold a version of reality within our minds, using abstract patterns to model real world behavior.[30] According to Deacon, the capacity for symbolic reasoning is our defining attribute, having co-evolved with language over a period of about two million years.

It is hard to imagine anything more deeply personal and individualistic than one's mental representation of the world, so it is bold as to suggest that wherever sentience emerges under natural [selec]tion, even if it occurs on some distant planet, we can expect religion [as w]ell.

[T]he Stark/Bainbridge understanding of religion is based upon what [they] call *compensators*. They define a compensator as a sort of IOU, or [sta]nd-in for some real reward that humans might seek. The promise of an [aft]erlife, for example, might be a compensator in the absence of any [ot]her available methods to evade death. It turns out, perhaps not [su]rprisingly, that humans prefer real rewards to compensators whenever [p]ossible.

The trouble is, of course, that the real rewards that humans seek are not always to be had. Stark and Bainbridge describe how in the 1930s, fundamentalist religion offered hope to deprived communities in Appalachia. Understanding that "material luxuries" were beyond their reach, "they defined these things as sinful and accepted the compensatory belief" that deprivation in this life would bring even greater rewards in the next. But later, when economic conditions improved for these groups, they quickly took advantage of their new circumstances and availed of the luxuries they had previously rebuffed.[36]

Stark and Bainbridge define religion as referring to "systems of general compensators based on supernatural assumptions."[37] An important aspect of their model is that it portrays religious belief as an essentially rational response, meaning that religion arose out of a need to make sense of things. In the words of anthropologist Melford Spiro:

> [Religious beliefs] continue to be held because, in the absence of any other explanations, they serve to explain, i.e., to account for, give meaning to, and structure, otherwise inexplicable, meaningless, and unstructured phenomena. They are assumed to be true because in the absence of competing beliefs, or disconfirming evidence, there is no reason to assume that they are false.[38]

In fact, as our ancestors became increasingly dependent upon rationalization for their survival, religious compensators were often the only means to satisfy their desires for elusive rewards and explanations. Death, in particular was a problem that required accommodation within

ironic that the ability to think in this way may h...
development of language, a highly cooperative and...
This characteristically human ability is what makes...
run simulations of reality in our minds, to examine...
outcomes, to revisit the past or to contemplate our futur...

Our expectations regarding the behavior of others a...
the fact that we know that they too can run these mental...
Deacon points out, we can more deeply empathize w...
humans "by extrapolation from one's own subjective ex...
ability which he laments "is also the source of the most de...
heinous of human practices," as it gives us the power to "...
suffering" that our actions may cause.[31] We are different...
species, says Tattersall, "in the fundamental sense that we, and...
are able to reflect on that difference."[32]

Whatever difficulty we may have in figuring out what our an...
were thinking, we can at least be certain that they *were* thinking...
was a way of thinking that gave rise to a relentless passion for...
expression, a strong desire to explain the world and their place withi...
and, as Tattersall puts it, "a curious reluctance to accept the inevital...
limitations of mundane human experience."[33]

With humans came the introduction of reason to an otherwise purposeless natural landscape. The record suggests that once rational, sentient humans appear on the scene, religious belief follows as a matter of course. In their landmark study *A Theory of Religion*, Stark and Bainbridge explain why religion was so inevitable.

Humans, with rational minds constantly on the lookout for cause and effect, cannot help but seek answers to such questions as: "'Does life have purpose? Why are we here? What can we hope? Is death the end? Why do we suffer? Does justice exist? Why did my child die? Why am I a slave?'" Once we begin to ask these kinds of questions, we lead ourselves to believe that events do, in fact, happen for a reason. "To seek the purpose of life is to demand that it have one."[34]

But if there is a purpose, then this implies the existence of some conscious entity possessing motives and intentions, a line of thinking that invariably takes us into the realm of the supernatural.[35] Indeed, given this appreciation of the relationship between sentience and religion, we might their worldview. Stark and Bainbridge note:

> When persons seek scarce but valuable rewards, they usually do not give up at the first sign of difficulty. Humans are persistent in pursuit of strongly desired rewards.[39]

With respect to the desire to evade death, this persistence is evident throughout our history. Impressive monuments such as Newgrange in Ireland, or the pyramids in Egypt, represent extraordinary human efforts to deal with the issue. The rituals associated with religious practices served another important function, namely the propagation of information that would otherwise have been lost when individuals died. In the absence of books, this process was vital to the development of human culture.

If religious compensators are required for scarce or unattainable rewards, then the corollary is that religion will be less important when humans seek more readily available resources, or when down-to-Earth matters are more easily understood. When there is an easier way to obtain rewards, say Stark and Bainbridge, "people tend not to seek them from the gods."[40]

We might then expect, with all the progress that has been made in science and technology, that there are many aspects of human existence where religion is no longer as important as it once was. A survey conducted by psychologist James H. Leuba in 1914 and again in 1933 showed that, without doubt, scientists were far less religious than the rest of the population.

He asked the top scientists of the day whether they believed in a god who responded to prayer, and in life after death; questions that he felt went to the heart of western religious belief. The answer, in 1933, was that more than eighty percent of top scientists were non-believers or agnostics.[41]

When the survey was repeated in the 1990s by Edward J. Larson and Larry Witham, the level of disbelief among National Academy of Science members who responded exceeded ninety percent.[42] The conclusion, it seems, is that the more scientific you are, the less likely you are to need religious compensators.

While religion is still alive and kicking for the rest of the population, its domain has shrunk considerably. Although there is no sign of a decline in the prevalence of religious belief, Stark and Bainbridge note that "religion is not the only institution capable of offering compensations for the rewards one cannot gain, or which do not exist."[43] Over the past few centuries, the promise of medical science, for example, has become an increasingly attractive compensator for many who would like to defer death and disease indefinitely.

Natural phenomena, for the most part, can be explained without recourse to religion, such that gods no longer have to intervene directly in say, cell division, evolution, or to adjust planetary orbits. It is perhaps for this reason that many Christian theologians have abandoned their practice of attributing everything in nature to some theistic design decision.

Religion is still our primary means of dealing with death, as evidenced by the ubiquity of prayers and religious rituals at funeral ceremonies, although its effectiveness may be diminishing. Aries describes how, during the course of the last century, we have distanced ourselves from death; it has become increasingly distasteful to the point where its reality is almost denied. In his view, the world was witnessing a major change in how death is viewed by society.

> Except for the death of statesmen, society has banished death. In the towns, there is no way of knowing that something has happened. . . . Society no longer observes a pause; the disappearance of an individual no longer affects its continuity. Everything in town goes on as if nobody died anymore.[44]

The same observations had been made by Geoffrey Gorer, who, in the 1960s, was the first to perform an in-depth study of the rejection of mourning in western society. In his essay, *The Pornography of Death*, Gorer described how death had become as unmentionable as sex was during the Victorian era. He showed that, whereas people used to be more familiar with death, improved public health and medicine now meant fewer fatalities among the younger population, so that a death in the family had become a relatively uncommon event.

> Whereas copulation has become more and more 'mentionable', particularly in the Anglo-Saxon societies, death has become more and more 'unmentionable' as a natural process.[45]

Another reason for the banishment of death, according to Gorer, is that without a belief in the certainty of an afterlife, "natural death and physical decomposition have become too horrible to contemplate or to discuss."[46] Traditionally, the only assistance people had in coming to terms with death was derived from their religious beliefs, and Gorer showed that, at least in England, these beliefs were wavering. His research, conducted in Britain in the 1950s and '60s, indicated that most people only paid lip-service to religious authority.[47]

Gorer points out that although death is depicted in the violence that we see in action entertainment, the more natural death, the kind that happens all the time in our communities, is largely ignored. Nobody wants to talk about it. Sex and death are both traditionally taboo subjects, and both have historically been governed by religious doctrine. However, Gorer notes how sex has moved more effectively into the secular domain, with positive results. It is now more acceptable to discuss human sexuality, to recognize sexual urges as natural, and to accept that suffering may result if attempts are made to suppress them. Death on the other hand, is kept hidden. Gorer describes a personal encounter with this reality after the death of his brother.

> A couple of times I refused invitations to cocktail parties, explaining that I was mourning; the people who invited me responded to this statement with shocked embarrassment, as if I had voiced some appalling obscenity. Indeed, I got the impression that, had I stated that the invitation clashed with some esoteric debauchery I had arranged, I would have had understanding and jocular encouragement; as it was, the people whose invitations I had refused, educated and sophisticated though they were, mumbled and hurried away.[48]

In the United States, Gorer noticed the same trend in rejection of mourning, and conjectured that this might be because of the generally accepted requirement to "enjoy oneself . . . so that the right to the pursuit of happiness [had] been turned into an obligation."[49] Perhaps death did not fit well with the American way of life. This new attitude towards death is something that worried both Gorer and Aries. From the time of our most ancient human ancestors, each society has had some way of handling death that surely involved a set of tried and tested rituals. Without these rituals and the support of a social framework, we become inept in our response to death, often busying ourselves with meaningless activity in an attempt to bury our grief.

Aries shows that this modern denial of death was unheard of in the past. Our oldest historical records suggest that the usual approach to death was one of acceptance. "They were in no hurry to die," he says, but when death came it was treated as something normal, as part of the natural order of things.[50]

There are some fundamental problems with the approach of trying to distance ourselves from death and pretend that it doesn't exist, not least of which is the fact of its inevitability. The hospitalization of death in the twentieth century meant that the dying became even more alienated from society. Doctors were often unwilling to admit that their patients were dying, seeing death as a medical failure. This meant that patients were increasingly kept in isolation and ignorance regarding their condition. Aries notes:

> "Hospital personnel have defined an 'acceptable style of facing death' This is the death of the man who pretends that he is not going to die. He will be better at this deception if he does not know the truth himself."[51]

Perhaps, he suggests, since modern society hasn't been able to do anything about the "scandal" of death, we are reluctant to admit defeat; "we act as if it did not exist, and thus mercilessly force the bereaved to say nothing."[52]

In the United States, physician Elizabeth Kübler-Ross observed that the medical community focused solely upon the technicalities of keeping their patients alive, often without any regard for how they were feeling, or what their opinions might be. In her 1969 book, *On Death and Dying*, she suggested that it was time to accept the reality of death, and to work toward achieving a better understanding of the process of dying, so that the terminally ill could be treated with more humanity and compassion, instead of just being regarded as objects whose lives needed to be prolonged at all costs, no matter how diminished or pathetic those lives became.[53]

Thanks to improved medical techniques, our ability to patch people up and keep them alive means that there are now more patients left incapacitated and helpless, their life functions supported by hospital machines. For many, this prospect is more frightening than death itself, as they contemplate the extra burden it might place upon on their families and friends. As the following quote illustrates, quality of life can sometimes be just as important to an individual as staying alive.

> I've had a full life. I've been fortunate. I don't want to go through a painful end. I don't want to be a problem to society or my family. I just hope that a truck hits me. I don't want to go to a hospital for months, and I hope eventually our society and our institutions will preclude that sort of thing. The church has been preaching the beauty of heaven for centuries but nobody seems to want to go there.[54]

Kübler-Ross reminds us that in times past, "suffering was more common," and religious beliefs were stronger in order to give meaning to the suffering. Childbirth, for example, was once a "long and painful event," whereas today it is possible to avoid much of this pain through the use of anesthesia. There is no longer any reason to suffer, or to believe that we will be compensated for this suffering in a life after death. "Suffering has lost its meaning."[55]

Paradoxically, she says, religious faith in an afterlife has declined while at the same time society abhors the reality of death. Of course, if we can avoid thinking about our eventual fate, then there is no pressing need to give any serious thought to the promise of immortality. But as we get older, it may become more difficult for us to ignore the inevitable. In 1968, Rodney Stark showed that belief in the afterlife increases with age.[56]

A British study conducted between 1965 and 1972 showed that eighty four percent of dying patients believed in the possibility of life after death, compared to thirty three percent of a control group.[57] It seems that the closer we get to death, the more likely we are to believe in the hereafter. Interestingly, although Kübler-Ross advocated an acceptance of death,[58] she also developed a keen interest in mysticism and near-death experiences, and became convinced that death was just a transition from this life to the next.[59]

The message, it seems, is that death is not something which is easy to accept. Aries, despite his account of how people were more accepting of death in the past, admits that they were never overly eager to embrace it. Historically, ritual and religious belief may have been more effective at integrating death into society, and providing the social constructs that allowed people to deal with their bereavement collectively, as a caring community. Nevertheless, death was never hugely popular. Sudden death in particular was greeted with shock – "a strange and monstrous thing that nobody dared talk about."[60]

When I read about the Cro-Magnon burial rituals, I sometimes feel sorry for people who thought and felt as we do, living in the darkest of times with so little technology, no knowledge of the biochemical processes of life and death, no real understanding of their place in the world, and no way to fend off death except through their myths and expressive rituals – a brave but somewhat futile stance against fate.

Yet, it may be that our circumstances are not much better. We live in a world that may be less meaningful to us than the world of our ancestors was to them; we are still subject to our own mortality, and the effectiveness of our rituals and beliefs has weakened. In fact, we do not know if our modern way of life will prove as durable in the long term as was the old hunter-gatherer lifestyle. Perhaps instead of pitying them, we should envy those ancient communities. For them death was expected and familiar, whereas for us it is something we try not to talk about.

In the decades since Kübler-Ross gave us the five stages of dealing with death, things have improved somewhat for the dying. Her work helped the hospice movement to become established in the US, and allowed many physicians to become better skilled at handling the terminally ill, learning to acknowledge the necessity and normality of death, instead of ignoring it. For Kübler-Ross and others like her, it was incredible that such an important subject as is our own mortality should be one that we are so willing to collectively ignore. Yet most of us would probably subscribe to the philosophy of the character Lister, from the BBC comedy show, *Red Dwarf*:

> Yeah, well, everyone dies. You're born, and you die. The bit in the middle's called life, and that's still to come.[61]

Like Lister, we can, for the most part, preoccupy ourselves with the rituals of modern life, which are so engaging and reassuring that we forget about our impending demise. Happily for most of us, death is more of an eventuality than something immediate. We can postpone worrying about such things and concentrate instead on everyday existence. That, by and large, is our approach; since we can do nothing about death, we either shelve the idea for later, or leave it in the realm of the gods. Religion is still a strong element in human culture, along with the belief that death is not really the end of things. Nevertheless, Kübler-Ross found that the majority of those facing death have something less than total belief; "not enough," she says, "to relieve them of conflict and fear."[62]

Hardly anyone embraces the idea of death if there's a realistic chance of avoiding it. The strong possibility that it might mean an end to our existence is disincentive enough. Perhaps it is for this reason that Americans invest so heavily in health care - to find new ways to stay alive. Our goal, it seems, is to live forever, or die trying. Since we are nowhere close to being able to live forever, it seems more likely that we will, in fact, die trying. But in doing so, will we destroy our quality of life? Aries, Kübler-Ross and Gorer seem to think so. Should we, as Kübler-Ross suggests, try to accept death as something natural and normal? Will we then, like her, end up embracing mysticism and a belief in everlasting life? Is death, as Gorer suggests, just too horrible to contemplate without a belief in immortality?

The difficulty with the concept of an afterlife is that, as a religious compensator, it is less than satisfactory for the scientifically minded. According to Stark and Bainbridge, science is the process by which "good explanations" can be obtained and "objectively evaluated," thus providing humans with real rewards rather than compensators. "Science erodes compensators," and "Bereft of its compensators, religion is nothing," they say.[63] But while we may prefer the truth, it is not always as satisfying. We render the compensators ineffective, but have nothing to put in their place. What does science have to say about immortality? And is there a way to rationalize death without resorting to the supernatural? Is there a new way to embrace death?

In trying to answer these questions we must recognize that human culture and technology continue to evolve at a breathtaking pace. How will the need to accept death be reconciled with the fight for life? Clearly these two goals are at odds with one another. Since Kübler-Ross began her crusade, our culture has been infused with a new respect for death, and a feeling that living too long is unnatural. Yet the fight-for-life approach continues to give rise to medical machinery and techniques to keep people alive longer. Which will prevail?

When I remember how as a child, I leveled these kinds of question at my parents, I am always impressed when I remember their resourcefulness in coming up with different answers for me. For the most part, they were patient, and tried to address my childhood concerns as best they could. Questions about death and dying were ones that I could tell they had difficulty with. So I turned to Granddad Waine, who was older and therefore wiser.

"Live every day as if it's your last, because some day you will be right", he used to say.

I thought it was funny how old people always want to remind us of the certainty of death. By the age of six, I had already taken a militant stance against the grim reaper. In response to my "what happens next?" question, Granddad first delivered the standard "you go to heaven" response, but perhaps exasperated by my continuous probing, he eventually thought of another solution.

"Look, when I'm dead, if there is any way possible I will come back and let you know what it is like", he promised.

Granddad Waine was an intelligent and well intentioned man, but he died from cardiovascular problems at the comparatively youthful age of sixty one, and he never made it back. Somehow, it did not surprise me, for the look of terrible anguish on the tear-streaked faces of my relatives seemed to suggest that Granddad was gone for good.

As religious as Granddad Waine was, even he, it seemed, could not rule out the possibility that death was nothing more than the beginning of an endless, dreamless sleep – a concept that, to my eleven year old mind, was intolerable. I was just emerging as a conscious being, surrounded by a loving family and the richness and potential of human society, only to learn that all of this, including my own thoughts and feelings about the world, might eventually be consigned to oblivion.

For me, there was no question more urgent, no issue more pressing. In the world of adults, these matters were addressed with Christianity. The answer, I was told, was to be a good Christian, and my salvation would be assured. But the explanations for how this might happen seemed to me, a little half-baked, and included an unhealthy dose of uncertainty. For this, the adults prescribed an equal measure of faith – a solution which I had some difficulty with. To rely upon doubtful doctrines as our only means of handling critical problems seemed to me a foolhardy approach. There had to be a better way.

So, with Granddad's advice in mind, I began my quest to deal with the problem of death and dying. I turned to books, but the question was, being only eleven years old and without Granddad's help, which books to read? In my last discussion with him on the subject, he told me about an ethereal thing called a soul, the part of a human being which survives beyond death. I pictured it as a sort of semi-transparent white cloud. The difficulty here, as I explained to Granddad, was consciousness. As I understood it, there was sufficient evidence to suggest that our memories and feelings were produced by electro-chemical happenings within our brains.

"So if the brain is destroyed, or decomposes, is the conscious mind lost, or can it escape with the soul?"

"What does it matter?" he responded. "Your soul will survive, and that's what's important."

"Well, no" I insisted. "If my consciousness is gone, then why should I care whether or not my soul survives? I won't know anything about it."

I remember he sighed, and was either unable or unwilling to go any further with this line of questioning.

"Perhaps," he said, "when you are older you can figure out all the answers."

When he died, the following occurred to me: If he was right, and there is a place called heaven, with a god caretaker, then there was nothing to worry about and Granddad was alright. If not, then Granddad was gone, and it was all up to me to figure out what to do. I was an eleven year old Gilgamesh, demanding to know how to overcome death.

"Must I die too? Must I be as lifeless as [Granddad]?"[64]

This was a difficult problem, and I was afraid that the answer might be beyond my reach, but in the end the subject was too compelling for me to ignore. Nothing commands quite the same level of attention as that which is quite literally the matter of life and death. Nothing is more relevant to all of us, from both personal and societal standpoints. And yet, as we have seen, it is an area that is often avoided and left draped in mystery and ambiguity, or given vague and unsatisfactory consideration. Like Gilgamesh, I was not content to leave the matter alone. If he could begin his quest anew, I wondered, where would it lead?

2

⌘

An Unpalatable Fate

At the turn of the century, astronomer Seth Shostak came to our local Barnes and Noble store to sign copies of his latest book about extraterrestrial intelligence. He described, among other things, what might happen if we were to make contact with an alien race, and one of his more interesting speculations was that the aliens would surely have discovered the cure for death. It was unusual, I thought, for a scientist to openly admit that death was a concern and to denigrate it as a kind of disease that would eventually be done away with.

I was reminded of Shostak's comment during a 2004 Wonderfest presentation at UC Berkeley entitled "Is there a Fountain of Youth for the Brain?" In the discussion that followed, the moderator told us that astronomer Dan Werthimer had made a similar pronouncement, lamenting that his was the last generation to die. Perhaps, I thought, astronomers had some particular aversion to death.

One of the speakers, Dale Bredesen, President and CEO of the Buck Institute for Age Research, was skeptical of Werthimer's claim that we might be close to eradicating death, while his co-speaker, Meredith Halks-Miller, Head of Pharmacopathology at Berlex Biosciences, wondered why anyone would want to.

"Death is what makes life significant, right?"[1]

Her question met with comparative silence. Maybe the audience, interested as they were in the promise of a fountain of youth, did not entirely agree with her on that point.

Perhaps it was not by coincidence that both Shostak and Werthimer were involved in the search for extraterrestrial intelligence (SETI), a concerted effort to detect the presence of galactic societies that communicate over the vast distances of space using radio waves. Humans first developed this capability less than a century ago, which is not a long time when compared with an estimated age of about 13.6 billion years for our galaxy. Unless there are communicative aliens sprouting up all over the place, our chances of detecting them depend heavily upon the assumption that the lifespan of a technological civilization can be measured on a scale that would make them, for the most part, truly ancient by our standards. One might suppose, therefore, that these SETI astronomers have at some length considered the nature of long-lived societies, and one of their conclusions, it would seem, is that death is a problem that invariably gets solved.

After the lecture I wandered over to Memorial Glade, settled down under a tree with my cheese and tomato sandwich and thought about Halks-Miller's comment. How did we feel about death? How should we feel about it? On the one hand, there is a modicum of political correctness in the view that death has value, that it endows life with meaning. We can acknowledge with fairly certain consensus that there is little point in questioning death, or in trying to argue against it. It happens to all of us, and as such it is a fact with which we have to live. Yet I am reasonably sure that relatively few of my contemporaries would feel unfairly diminished by death's absence. Despite its inevitability, it is more common for us to ignore the matter completely; to avoid talking about it. As author P.D. James observed, such is our discomfort with death that it has become a taboo subject.

> Huge isn't it? And much more so in our century than it was for the Victorians. Of course the Victorians had to get used to it far more than we do, I mean it's appalling to think that in those days you could expect, you know, probably a third of your children to die before they grew up. But they took it in their stride because I mean they had all sorts of formal procedures for mourning, didn't they? And they liked to be at death beds, and people had their last words. Now, we just tidy it all away and we're all expected to die in hospitals and give no offense

to anyone. And it has become the great unmentionable. It's interesting that sex we can discuss very fully, in all its manifestations and all its oddities even, but death, no – we do shrink from it. In fact as people say, in some societies it's almost regarded as an optional extra.[2]

If death is indeed what makes life significant, why then are we so slow to accept it? The record shows that humans have persistently tried to evade the inevitable. The earliest sign of reluctance to accept the natural order of things is found in the presence of grave goods at prehistoric burial sites. This is a testimony to the antiquity of belief in an afterlife, a belief that did not disappear but instead became woven into the cultural fabric of our most sophisticated civilizations.

Nowhere was this belief more dramatically evident than in ancient Egypt, with the civilization that first settled the Nile valley some 5,200 years ago and which lasted for almost three millennia. The secret to the prosperity and longevity of the Egyptian settlements lay with the Nile itself. Every year, the great river would flood its banks bringing water and nutrients to the surrounding farm land. The land was kept fertile by the cyclical flooding, which meant that there was a constant and reliable food source to support the local population.

This, in turn, allowed for the growth of resource rich communities throughout the region, and an unprecedented concentration of wealth. The narrow strip of fertile land along the river was shielded by eastern and western deserts, and by the Mediterranean in the north, which meant that the inhabitants of the Nile valley enjoyed relative isolation and protection from outsiders. The result was prosperity, economic and political stability, and a way of life that saw little change as the centuries passed.

Beyond the point where immediate needs were met, the pace of technological and cultural change was slow. In the absence of pressing imperatives, constancy prevailed, the year-round sunshine yet another reinforcement of the sense of permanence and continuity that pervaded Egyptian life. In such a setting, it seems entirely appropriate that people would believe in the persistence of life beyond the grave. Any other interpretation of death would have seemed out of place, in a world where everything stayed the same.

Every society looks to its past for reassurance, for legitimacy. If we in the United States can derive a sense of identity and privilege from the artifacts and monuments of our several hundred year history, how much more privileged must the Egyptians have felt? When the Egyptian king Seti I built the Great Temple of Abydos about three thousand years ago, the city was already more than two thousand years old; old enough perhaps, to give the citizens of Abydos a sense of eternity. Perhaps not surprisingly then, Egyptians imagined the afterlife as being not so different from their everyday existence.

In their new spirit lives, the waters of their beloved Nile would continue to bring them life, as they tended the fertile lands forever. Hopefully, of course, life would get easier after death, and there would be servants to help with the work. People were buried with miniature carvings of servants, farm implements and other treasures that would be of assistance to them in their new lives. However naively optimistic this might have been, it was of course far better than the practice of having your actual servants buried with you, as some kings were known to do.

Death, it seems, held much fascination for the Egyptians, and was a significant driver of economic activity. Much of what we know about Egyptian life comes from what was found in their tombs, and there is little doubt that being adequately prepared for death was of high importance in life. Historian Paul Johnson notes:

> An Egyptian spent his adult years planning for eternity just as, today, the insurance companies tell us we should plan for retirement.[3]

As in Judeo-Christian tradition, Egyptians were supposed to live in honest accordance with a set of established moral principles in order to attain eternal salvation. They believed that on the Day of Judgment, they would stand before the gods and be forced to account for their misgivings in life. Anyone wishing to enter through the gates of immortality needed first to answer a series of questions to the satisfaction of Osiris, the god-in-chief presiding over such matters. With so much riding upon getting these answers correct, many among the more influential citizenry understandably felt that it would helpful to know in advance what the questions would be, rather than leave things to chance.

A collection of scrolls called the 'Book of the Dead,' was devised, detailing the questions that would be asked on the Day of Judgment, along with the responses that Osiris would expect.

Thus the Day of Judgment became little more than a formality; as long as you carried a copy of the Book of the Dead, you were pretty much guaranteed to pass safely through to the other side. Short versions of the scrolls became available, often less elaborate and produced in bulk so that they could be sold more cheaply to the populace, a practice not too dissimilar to the selling of indulgences by the Vatican in more recent times. The souls of the faithful had a rubber-stamped assurance of a place in the hereafter.

Death became big business, with scrolls, carvings, and miniature statuettes called shabtis manufactured and sold in the thousands. Other accessories were available for the well-heeled, such as scroll holders, or elaborate stone coffins which sometimes had selected poems from the Book of the Dead inscribed on them. By far the most expensive necessity was the tomb itself. Burial was important for the dead, because Egyptians believed that the corpse served as a receptacle for the eternal soul. Without a body to return to, the poor soul would have no home, and would be denied its new life.

It was clear that unless a dead body was protected in some way, it would decompose and so the many generations of Egyptian civilization saw continuous improvement and innovation in tomb design and cadaver preservation. To prevent decomposition, bodies were first treated with a drying agent called natron. They were then wrapped in linens, often soaked with resin to protect against moisture. Internal organs were removed before the wrapping began, and these were often placed in a separate bowl which in some tombs was housed in a special alcove next to the coffin.

The mummification process, as it was called, became perfected to such an extent that the bodies of thousands of Egyptians remained well preserved for millennia. During the mummy craze in the nineteenth century, mummies were unearthed by the thousands. Most were ground into powder that was sold as a remedy for countless ailments. Some ended up at mummy parties in the homes of the European elite, where the highlight of the evening was the unwrapping of a mummy to entertain the guests. What a sad fate for the mummies; carefully preserved through the millennia, only to be pulled apart for the amusement of inebriated aristocrats.

Figure 1. Pyramid of Khafra, at Giza. Photo by Olaf Tausch.[4]

For the Egyptians it was important that each family member receive a decent burial. Cost was often an issue. In addition to the procurement of a tomb, and the various obligatory funerary artifacts, tombs often required ongoing maintenance and security to prevent break-ins. For the same reasons that many buy life insurance today, it was not uncommon for a farmer to set aside a plot of land, the income from which could later be used to pay the funeral costs for a son or daughter.

Considering all the attention given to funeral preparations and the extent of economic activity devoted to the subject, it might seem reasonable to conclude that Egyptians were overly preoccupied with death. From what remains of their civilization, one could almost say that death for them was a way of life. Many Egyptologists however would disagree, and point out that what we see is not so much a fascination with death as a passion for life. These inhabitants of one of the most fruitful places on Earth were so addicted to life, so impressed with a sense of permanence and prosperity that they couldn't see themselves being torn away from it.

The Egyptians did not look upon death as an end to existence, nor did they contemplate the start of some bizarre and glorious new beginning. Instead they looked for a continuation of the life they already enjoyed. Clearly they did not wish for their lives to end. In that sense, it is remarkable that the first truly long-lived and expansive civilization on the planet became humanity's greatest anti-death demonstration – our boldest effort to avoid annihilation.

It's ironic that a civilization so preoccupied with death should have difficulty accepting death at face value, and instead hold fast to the promise of continued life beyond the grave. When an individual dies, the body becomes lifeless and inanimate, and is noticeably devoid of the character and personality it once housed. Surely it makes more sense to assume that, once dead, a person is gone for good, rather than hypothesizing the existence of an afterlife in the west, supported by a cast of mythical characters. As historian Arnold Toynbee puts it:

> It seems reasonable to infer that, when the body dies and disintegrates, the personality, which instantly disappears, is annihilated simultaneously.[5]

The likelihood that death brings nothing but an endless, dreamless sleep was not lost on the ancient world. It is a view most often associated with the Greek philosopher Epicurus who lived in the fourth century BCE:
> So that the man speaks but idly who says that he fears death not because it will be painful when it comes, but because it is painful in anticipation. For that which gives no trouble when it comes, is but an empty pain in anticipation. So death, the most terrifying of ills, is nothing to us, since so long as we exist, death is not with us; but when death comes, then we do not exist. It does not concern either the living or the dead, since for the former it is not, and the latter are no more.[6]

The finer points of Epicurean philosophy were expounded two centuries later by the Roman poet Lucretius. He notes that consciousness depends upon sensory input provided by our eyes and ears, the physical apparatus of our bodies, and is therefore unlikely to persist in disembodied spirits. "Mind and memory", which are often seen to deteriorate with age, eventually fade to nothingness when we die.[7]

> Now, if the mind is immortal and can feel
> When parted from the body, we must assume
> It has the five senses. Only in this way
> Can we imagine the spirits of the dead
> Go wandering in Hades. Painters and poets
> Have always shown us spirits endowed with senses.
> But what do you think? Can a spirit without body
> Have eyes or nose or hand or tongue, and can
> The ears hear by themselves without a body?[8]

When you are dead you won't feel any pain. In fact, you won't be aware of anything, since you won't exist. And why should this worry us, asks Lucretius, since neither did we exist in the eternity that preceded our birth?

> Look back upon the ages of time past
> Eternal, before we were born, and see
> That they have been nothing to us, nothing at all.
> This is the mirror nature holds for us
> To show the face of time to come, when we
> At last are dead. Is there in this for us
> Anything horrible? Is there anything sad?
> Is it not more free from care than any sleep?[9]

And their religious beliefs notwithstanding, it was obvious even to some Egyptians that the promise of eternal life might not be worth the treasure invested in it, and that death may simply be the end of things. The harpers' songs from ancient Egypt reveal such skepticism about the reality of an afterlife:

> Gladden your heart, let your heart forget!
> it is good for you to follow your heart as long as you exist.

> Put myrrh on your head,
> clothe yourself in white linen,
> anoint yourself with genuine oil of the divine cult,
> increase your happiness, let your heart not weary of it!
>
> Follow your heart in the company of your beauty,
> do your things on earth, do not upset your heart,
> until that day of mourning comes to you.
> The "weary of heart" does not hear their cries,
> and their mourning does not bring the heart of a man back from the netherworld.[10]

It is a message that echoes through the ages from the time of Gilgamesh, and that still holds much sway in modern times; "Eat, drink and be merry, for tomorrow we die."[11] In fact, there is now an even greater body of knowledge to support the Epicurean viewpoint. For example, the more we learn about the human brain as the repository for personality and awareness, the more it underscores the real relationship between physical bodies, or brains, and the essence of human existence and experience.

It appears that our thoughts and memories are the result of a cacophony of electrical signals cascading between sets of nerve cells, or neurons. This is the basis of human personality, and however spiritual we may become, however ethereal our minds may seem, we are shackled by our brains to the realities of the physical world. Our thoughts and actions can be directly affected by drugs or electrical stimulus. When brain physiology is damaged due to injury or disease, remarkable changes can occur in a patient's personality or state of mind, as described by Oliver Sacks in his book *The Man Who Mistook His Wife for a Hat*. The slow, progressive deterioration brought about by Alzheimer's disease can cause people to literally lose their minds. It seems unlikely that a mind could survive the demise of its body. Our new exploration of the close link between mind and body is taking us closer to what Sacks calls a "'neurology of identity', for it deals with the neural foundations of the self, the age-old problem of mind and brain."[12]

Having removed the brain, as was the customary practice in

mummification, it was unreasonable to expect that the remaining carcass could accommodate any of the personality or conscious content of the individual. As a result, says Epicurus, when you die, you die. You won't feel anything, so why worry about it? This is exactly the line of reasoning my Dad used with me thirty years ago. Although the logic is hard to dispute, I still feel that it misses the point. It sounds very much like the old advice attributed to Harry "Breaker" Morant: "Live every day as if it were going to be your last, for one day you're sure to be right."[13] There may have been a time when it was possible to live like that, but if I were to live every day as though it were my last, I wouldn't be able to pay the mortgage next month.

Planning for the future is one of the traits that distinguish us from the other animals, and this practice of thinking ahead is what gets us cogitating about our ultimate departure. My Mom's brother was given a year to live, after being diagnosed with advanced small cell lung cancer. On one of their morning walks in the park, she asked him if there was ever a moment in which he could put it out of his mind – distract himself from his fate. "It's always there," he answered.

Some may argue that life needs to be finite; otherwise there'd be no motivation to get anything done. And yet it's hard to seize the day when faced with oblivion. If the assumption from Epicurus is that people worry about how they will feel when they are dead, rather the issue may instead be that people do not wish to be dead in the first place.

The Egyptian experience dramatically illustrates the promise of existence beyond death as an undeniably powerful component in the development of human culture. For as long as there have been humans, there has been a belief in the perseverance of the human soul. For the philosopher Plato, his respect for the dignity of the mind, along with his passion for reason left no other conclusion possible. The soul must exist. Although the Epicurean explanation of death may make more sense, Toynbee notes that it has never enjoyed popular support.

> This general rejection of the hypothesis of extinction might be discounted as being inspired by self-conceit or by an unwillingness to face an unpalatable truth.[14]

It is prudent to be cautious when it comes to death. All animals have a strong desire to remain alive, and humans are no exception. As certain as Plato was, and as powerful as religious systems have become, there have always been individuals who were unable to hold fast to a belief in the hereafter. If we cannot be sure that we will survive death, might there be a way to avoid it entirely?

The idea that we might cheat death is perhaps as old as the human imagination. In what may be the oldest story ever written, Gilgamesh, a king tormented by thoughts of death embarks on a mission to discover the path to immortality. His journey was filled with strange encounters and bizarre happenings. When most of the world was terra incognita, there was apparently no limit to the exoticism that an intrepid explorer might expect to find.

According to the Greeks there was a land in the distant north, where people lived extraordinarily long lives, and seldom became ill. Inhabitants of the southernmost extremities of the Greek world were similarly endowed with longevity and healthiness, in addition to being extremely tall. Travelers' tales were embellished with all manner of fanciful creatures with unusual traits. Lucian Boia recounts the story of a Greek explorer who landed on a remote island in the Indian Ocean where he encountered a race of people who seemed immune from the effects of old age.[15] Explorers carried their preconceptions to distant lands, and new discoveries were often seen through the lens of mythology.

When Juan Ponce de Leon came upon the coast of Florida in 1513, he was said to be searching for a fountain of youth that was purported to be in the area.[i] According to local mythology, the fountain was located on the fabled island of Bimini, and its waters had the power to reverse the effects of aging, restoring youthfulness and vigor. The concept would not have been considered too outlandish by sixteenth century standards. Indeed, even today, visitors to the Fountain of Youth Archaeological Park in St. Augustine still drink from what is advertised as the real fountain, presumably hoping that there might be some truth to the story.

Miraculous fountains also play a role in the *Alexander Romance*, an

---

[i] According to an account by Gonzalo Fernandez de Oviedo, in his *General History of the Indies*, Ponce de Leon had heard about the fountain and its powers of rejuvenation from some of the local islanders. It is now generally believed that the story is untrue.

epic poem describing the exploits of Alexander the Great. In a twelfth century version of the story, Alexander hears of a fountain that bestows immortality to those that bathe in it. Unfortunately the effect can only be achieved once per year, and when Alexander reaches the fountain, he is furious to find that another man has already availed himself of the water's healing powers, thus rendering them useless for another year. For his punishment, the offender is incased in a pillar of stone – a frightening fate for an immortal.[16]

There are innumerable variations on the fountain of youth theme, encapsulated in the stories and traditions of the many cultures around the world. The famous Achilles was empowered by magical waters at a very young age, when his mother held him by his heel and dipped him into the river Styx. He became invulnerable, except for the area around his heel which, sadly for him, led to his defeat in the Trojan War.

In my early school days in Ireland, we learned of a mythical land in the west, called Tír na nÓg (Land of Youth), where people never grow old or die. A young warrior named Oisín was lured away to this land by an attractive female on a white steed. They lived a blissful life for many years, and retained their youthful energy and appearance. Eventually Oisín returned to visit his homeland, and as he rode through the Irish countryside, he saw some men trying to remove a huge boulder from the road. He had been warned not to dismount from his enchanted horse, but as he stooped down to assist the men, the stirrup leather snapped and he fell to the ground, whereupon he instantly transformed into an old man, losing the gift of eternal youth. I remember wondering what possessed Oisín to risk everything, just to help move a stupid rock.

In the absence of miracle fountains or mythical horses, people have shown remarkable resourcefulness in coming up with ways to stave off the inevitable. Sleeping with virgins was one such prescription for longevity that was taken seriously, especially by older men. Sex wasn't necessarily a vital element in the arrangement, for it was believed that a man could be revitalized by just the warmth from a woman's body. The biblical King David is thought to have employed a young woman for this purpose – to warm his body at night. The seventeenth century Dutch physician Herman Boerhaave prescribed similar treatment for an old burgomaster, although perhaps due to the colder climate of the Netherlands, he upped the dosage and had the old man lie between two maidens, for extra warmth.[17]

Taoists held that by having sex you could extend your lifespan, by absorbing a life-giving sexual energy, or essence emitted from your partner at the moment of climax. It was important, however, to avoid achieving orgasm yourself, so as not to waste any of this vital energy.

Among the more bizarre procedures for life prolongation was that of the French surgeon Serge Voronoff who reckoned that transplanting chimpanzee testicles into human subjects was the way to go. Voronoff believed that the grafting technique would rejuvenate men by improving their sexual virility. Fellow Frenchman Alexandre Guéniot took a very different view, and felt that too much sex was detrimental to long life. Smoking, on the other hand, was considered by Guéniot to be relatively harmless.[18]

While the remedies and recommendations have changed over time, our supermarket shelves are still fairly replete with dubious products that claim to combat the effects of aging. Author Robin Marantz Henig described how Ana Aslan became "one of the richest women in Romania, all from the sales of Gerovital - which turned out to be nothing more than simple Novocain."[19] Unscrupulous fraudsters still abound, ready to take advantage of our desire to stay young. A report in the March 1994 edition of *FDA Consumer* noted:

> Some anti-aging products are also promoted to either prevent or treat Alzheimer's disease. According to JoAnn McConnell, Ph.D., of the Alzheimer's Association, "so-called new 'cures' for Alzheimer's surface constantly."
>
> But there are no cures, which may cause Alzheimer's patients and their families to be susceptible to products holding out false hope.[20]

However, on the whole we have benefitted greatly from the advances in medicine and improvements in public health which have flourished during the last few centuries. People are living longer, and enjoy more active lifestyles well into their later years. By tackling such diseases as Alzheimer's and Parkinson's, the efforts of scientists such as Halks-Miller and Bredesen may help to extend lives still further. Unfortunately this may bring a new set of problems, as more of us begin to encounter the degenerative conditions that disproportionately affect our older population.

Medical breakthroughs are now eagerly anticipated; each new discovery is heralded on nightly newscasts as the long awaited elixir of life. The discovery of telomeres, for example, was greeted with proclamations that immortality would soon be within our reach. Telomeres are the short repetitive DNA stubs that are found at the ends of chromosomes, somewhat like the aglets on shoelaces.

Each time the DNA is copied, during cell division, the telomere gets a little shorter. This is because the copying mechanism is unable to reach all the way to the end of the DNA strand, so a little piece of DNA gets chopped off at the end. In this way, the telomere acts as a buffer, preventing useful DNA from being snipped away.

Figure 2. Lucas Cranach's famous depiction of a fountain of youth. The elderly and decrepit enter on the left and emerge rejuvenated and invigorated on the right.[21]

Cell division is the method by which old cells replenish themselves, but eventually, when we reach a certain age, the telomeres have dwindled away to the point where the copying process is halted, and no further division takes place. Perhaps if we could make our telomeres a little longer, we could prevent this cellular replenishment from shutting down – hence the promise of rejuvenation and immortality.

Immortal cells may sound like a good idea, until you recognize the role that they play in cancers. It turns out that cancer cells use an enzyme called telomerase which allows them to prevent their telomeres from getting shorter, making the cells effectively immortal. This is a fairly typical pattern in bioscience research. Almost as soon as a promising new field opens up, a whole new set of complications are also realized, due to the complexity of interactions at the cellular level.

Dealing with age-related illnesses is likely to be especially challenging, because in a very real sense it represents a struggle against nature. Diseases that affect the young can sometimes be remedied by making small corrections in nature's programming; compensating for a genetic flaw by supplying a missing enzyme, for example. Over time, we can begin to put an instruction manual together, detailing how to fix the various coding errors that cause disease.

The afflictions of the elderly, however, are a different matter. Not only are we missing an instruction manual, we also have to cope with the fact that nature never supplied any programming to keep us alive as we grow old. Unlike computer software, life's biological patterns and coded instructions were not compiled by an engineer.

Instead, our genetic code was generated via the process of natural selection; the design decisions being essentially governed by the breeding success of the organism. If any given change to the programming had a positive effect on your love life, then it was retained for the benefit of your descendants. As a result of the small improvement in your genes, your kids would presumably be more likely to survive and have offspring of their own. The important point to remember is that success, in this case, is defined as having more descendants. If you are still in the pink at age 250, but have no grandchildren, then your genetic material isn't likely to survive.

Thus natural selection was a very ineffective mechanism for designing bodies that would last. In engineering parlance, you might say that longevity was never on the list of requirements, and so was never factored into the design. It is possible, of course, that the proclivities that allow you to engage in profuse proliferation might also serve you well in your twilight years. But as the deliberations of natural selection worked their magic on the blueprint for humanity, the fate of the elderly was sadly neglected. Once the kids move out, nature's warranty expires, and our bodies begin to fall apart. This paints a rather gloomy picture as far as the subject of longevity is concerned, for as the human lifespan gets pushed beyond its natural limits, we may end up with host of new age-related disorders to manage. Trying to fix them all might seem like a game of *Whac-a-Mole*.

My granddad was a big proponent of modern medicine, especially vaccinations for babies. When I arrived on the planet, he encouraged my mother to get the full battery of immunizations for me, including the one for tuberculosis, which was still a concern in Ireland at the time. I remember a conversation with Granddad, sometime when I was little, in which he explained the concept of immunization, and how we could be protected from dreadful afflictions, such as polio and tuberculosis. It was, I believe, the first time I became aware of the incredible potential of our science and technology. We could literally keep people from dying if we so chose. I felt fortunate to have been born at a time when life expectancy was on the rise and problems were being solved. By all accounts, I could expect to live longer than my ancestors and have fewer ailments to boot.

Now that I am old enough to understand statistics a little better, I am no longer as optimistic about my chances of seeing the twenty second century. This is because increasing life expectancy is all about those who might have died, but didn't. To get a better grasp on this, all you need to do is to pay a visit to one of those old graveyards from times past, and take a look at some of the dates on the tombstones. As author Bill Bryson observed while strolling through a cemetery in Peacham, Vermont, you'll notice that an unconscionable number of individuals died at a relatively early age.

> A small panel on the back said:
>
> Nathan H. died July 24 1852 AE. 4 Y'S 1 M'O.
>
> Joshua F. died July 31 1852 AE. 1 YR 11 M'S.
>
> Children of J. & C. Pitkin.
>
> What could it have been, I wondered, that carried off these two

> little brothers just a week apart? . . . Everywhere I looked there was disappointment and heartbreak recorded in the stones. . . . So many were so young. I became infected with an inexpressible melancholy as I wandered alone among these hundreds of stilled souls, the emptied lives, the row upon row of ended dreams. Such a sad place! I stood there in the mild October sunshine, feeling so sorry for all these luckless people and their lost lives, reflecting bleakly on mortality and on my own dear, cherished family so far away in England, and I thought, "Well, fuck this," and walked back down the hill to the car.[22]

In seventeenth century England, only one out of every three children could expect to live beyond the age of four.[23] That this comes as a shock to the modern reader is as good an indication as any that things have, mercifully, improved a lot since then. In fact, the reduction in childhood mortality rates was the single biggest reason that average life expectancy took an upward turn. Another major factor was a welcome decrease in the number of women who died during childbirth. Overall advances in public health have increased life expectancy by about twenty five years in the past century.

Americans, on average, now live into their seventies, whereas for the majority of humans who have ever lived, life expectancy at birth was only about twenty.[24] Further improvements are possible. If we could somehow cure all cancers, says epidemiologist Donald Redelmeier, this would increase life expectancy by another 3.5 years.[25] Eliminating heart disease could get us an extra twelve, according to Leonard Hayflick from UCSF.[26]

But the fact that life expectancy has risen, doesn't necessarily imply a lengthening of the human lifespan –the maximum length of time that any human could expect to live. Medical historian Gerald Gruman tells us that lifespan is estimated by statisticians to be about 110 years.

> Unlike life expectancy, the span of life does not seem to have increased noticeably during the course of history. In every era there seem to have been a few hardy individuals who lived beyond one hundred years. The average man may have died

prematurely at twenty or thirty, but, then as now, a tiny quota of long-lived persons somehow managed to survive to the extreme limit of the life span.[27]

So unfortunately, despite what I was led to believe, the human lifespan hasn't exactly been skyrocketing. In fact, it doesn't appear to have changed much at all over the millennia. If you approach the 110 year mark, then you can consider yourself old, no matter what century you are in. The search for miracle cures and fountains of youth has been largely unproductive, and medical science doesn't offer much in the immediate term as far as eternal life is concerned.

So where does that leave us? Religion is one option, of course, and it is the one that most of us choose, most of the time. Another option is to simply accept the inevitability of death, and give up trying to do anything about it; an approach that Gruman calls apologism. The apologist's conclusion is that a life spent worrying about death is a wasted life. It is better, instead, to make the most of what we have, and to accept that life is its own reward.

From the ancient story of Gilgamesh, through the philosophy of Epicurus, it is a pervasive wisdom which is hard to rebuff, and is prevalent even in our most futuristic fiction. In the TV adaptation of Ray Bradbury's *The Martian Chronicles,* a Martian ghost reveals to Col. John Wilder that the secret of life is life itself.

"There is no secret. Anyone with eyes can see the way to live."
"How?" asks Wilder.
"By watching life, observing nature and cooperating with it; making common cause with the process of existence."
"How?"
"By living life for itself, don't you see? Deriving pleasure from the gift of pure being."
"The gift of pure being," repeats Wilder, thoughtfully.
"Life is its own answer," says the Martian. "Accept it and enjoy it day by day."[28]

The Martian's advice seems very similar to that given by Siduri, the divine barmaid, to the great Gilgamesh. And in the movie, *Star Trek:*

Generations, starship captain Jean-Luc Picard admonishes a would-be immortal, telling him that "It's our mortality that defines us. . . . It's part of the truth of our existence."[29]

Of course Picard's viewpoint is not really that of a twenty fourth century space traveler. Rather it was intended to be politically correct enough to resonate with a mass audience of science fiction aficionados from the late twentieth century. Apologism carries a certain intellectual appeal, perhaps even an aura of sanguinity. Denial of death is surely a philosophy for fools; as futile a proposition as King Cnut's attempts to turn back the tide.

On the other hand, acceptance of death as a normal consequence of living is surely the mainstay of enterprising realists and progressives. Apologism is pragmatic - where would we find the space for everyone if people started refusing to die? As such, the apologist position also represents altruism toward future generations.

In the long term, however, this way of thinking is unlikely to hold sway. For the most part, we are truly comfortable in our acceptance of death, only when we've rationalized our fate as something other than the annihilation of the soul. What we learn from history is that we are defined, not by our mortality, as Picard would have us believe, but by our defiance of the realities of nature.

Among the species with which we share the planet, we alone can contemplate our personal demise, and we've never been comfortable with the outcome. Every culture has made some attempt to circumvent the dilemma of human fate, or to explain it away using mythology or religion. Our most compelling stories are those that suggest the possibility of escaping our eventual doom. In the New Testament, for example, Jesus talks explicitly about God's promise of eternal life, and this message of salvation was undoubtedly a factor in the spread and persistence of Christianity around the globe.

> For God so loved the world, that he gave his only begotten Son, that whosoever believeth in him should not perish, but have everlasting life.[30]

Listening to the radio while driving home from work one night, I heard someone who had phoned NPR wondering what he might do to prepare for the experience of death. I do not remember what the show was about, but it did remind me of something that Terrence Deacon elucidated in *The Symbolic Species: The Co-evolution of Language and the Brain*. It was apparent that the caller hadn't really grasped the fact death isn't actually experienced. Rather, it is the end of experience. For this reason, says psychologist Jesse Bering, "You will never know you have died,"[31] or as writer Johann Wolfgang von Goethe puts it: "Everyone carries the proof of his own immortality within himself."[32]

This sobering fact is not one that sits easy with the human psyche. Other animals are not burdened with ruminations about mortality. When the time comes, they simply die. As Deacon points out, we are maladapted in that regard; we are plagued with the ability to feel "a foreboding sense of fear, sorrow, and impending loss with respect to our own lives, as if looking back from an impossible future"[33] and yet we are ill-equipped deal with this ability. Deacon tells us that there are two common approaches to the problem – one is religion; the other is to avoid thinking about it, and to allow ourselves to be continually preoccupied with the distractions of life.

> What great efforts we exert trying to forget our future fate by submerging the constant angst with innumerable distractions, or trying to convince ourselves that the end isn't really what it seems by weaving marvelous alternative interpretations of what will happen in "the undiscovered country" on the other side of death.[34]

So a common theme for humanity is that we are forced by our nature to transcend the limitations we encounter; we stubbornly refuse to accept things the way they are – so we reimagine the world, reshaping it with our minds and our technology. Thus we endow our lives with meaning, through our faiths, our personal mantras and our hopes. These are also the ways in which we make death meaningful, so that we can view our mortality in a context that makes living worthwhile.

Historian James Burke explains how medieval monks carried this practice to an extreme. Although they are often credited with preserving literacy and knowledge through the dark ages, everything in nature was seen and understood through the lens of their religion.

Nature's true meaning was not made visible by God. Nothing, therefore, was what it seemed. The 'Book of Nature' was a cryptogram that had to be decoded by the faithful. . . . Red was both a colour, and a symbol of the blood of Christ.[35]

For the pragmatist, religious interpretations of death are often regarded as nothing more than wishful thinking. In my years as a volunteer presenter for The Planetary Society, I often noticed that some members of the audience became a little disquieted by the enormity of the universe, and the vast spans of time that were being discussed. Whether I was talking about the timeline of planetary evolution, or looking back at our home world from six billion kilometers away, I discovered that providing an astronomical perspective was an excellent way to diminish a sense of self-importance.

When the agnostic astronomer Carl Sagan was asked "How can we feel worthwhile?" he delivered the clever response "Do something worthwhile."[36] The difficulty with this response and with apologism in general, is that it isn't satisfying enough. An apologist will forever find it a challenge to provide any real comfort to those who are suffering from the realization that their lives are drawing to a close.

While visiting relatives on the east coast, my wife and I met with a family friend who was struggling through her final days. She was angry about what old age had done to her; angry about the loss of mobility, the confusion and forgetfulness, the pain and discomfort that she had to endure, and her impending death. The visit made an impression on me, because it may have been the first time that I'd heard an old person speak this way. I guess I had assumed that the elderly somehow manage to resign themselves to their lot, but here was a woman who didn't mind saying that dying sucked.

In fact, says author Catherine Johnson, it would be mistake to assume that all seniors are ready to die, since studies have shown that even those who are sick and disabled will often describe their condition as satisfactory.

> Studies by Joel Tsevat of the University of Cincinnati Medical Center found that 43 percent of his subjects in the worst physical condition and 51 percent of those with severe pain

described their quality of life as good.[37]

Where ingenuity and technology are applied to improve the quality of existence, it's usually because someone complained about the status quo. When it comes to the issue of aging and death, it would seem from our history that humanity has made its discontent pretty clear. From an anthropological perspective it might even be argued that death is an inappropriate proposition for our kind.

Mortality is a trait that we share with other animals, and all of us, through the cycle of life, are replaced by our offspring. But for us sentient beings, the normal course of nature is an especially cruel and degenerative cycle. We live in a symbolic world of our own making, quite apart from the physical world, and inaccessible to the minds of other creatures. Within this world, the incredible uniqueness of each individual underscores the preciousness of human life. So much is lost when a person dies, that death is an affront to us. How might the determined minds of future generations deal with this intolerable situation? What remedies might they bring to bear?

Bicentennial Man is a movie in which actor Robin Williams plays a robot, traumatized at the deathbed of a woman whom he has known and loved since she was a child. With his eyes resting upon her lifeless body, the grieving robot is dismayed by the finality of death. "That won't do," he says.[38]

So he embarks upon a series of steps to develop new replacements for failing human organs, along with a regimen of anti-aging drugs that would allow people to prolong their lives indefinitely. The movie then adopts a rather obvious deference to apologism. Faced with the prospect of a lonely existence among a race of mortals, the robot chooses instead to relinquish his own immortality in an effort to be recognized as human.

Of course, with the availability of medicines and techniques for life prolongation, it is more likely that the robot would have found company among a growing population of individuals with a preference for longevity. Over time, the attitudes of such individuals would prevail, since those with a different view would be relatively short-lived, and so in the long term few would agree that the robot's decision to die had rendered him more human.

Whether or not we solve the problem of death, it seems to me that the answer to the question we posed at the beginning of the chapter is no – death doesn't define us. Striving against it is what defines us. If humanity could be said to have a goal, it would be, as reflected in the words of anthropologist Ernest Becker, to conquer death.

> We admire most the courage to face death; we give such valor our highest and most constant adoration. . . . The great triumph of Easter is the joyful shout "Christ has risen!", an echo of the same joy that devotees of the mystery cults enacted at their ceremonies of the victory over death.[39]

It's not in our nature to accept defeat, even when presented with the most formidable opponent – inevitability. We are obstinate in the face of adversity, just like the duck in the movie *Babe*, who refused to accept that he was destined for the dinner table. When the cow told him to accept things as they were, he responded with "The way things are, stinks!"[40]

For as long as our species may survive, our path will be to change the way things are, to pursue our quest to improve the world for the benefit of humankind, and to challenge death's dominion through compassion and cleverness. We will continue to respond to that human imperative to push back the boundaries that nature has set for us, and to reimagine the human soul.

3

⌘

A Leap of Faith

My secondary school was run by an organization called the Christian Brothers, men who walked about in long black robes and who had a reputation, at least where I grew up, for being somewhat ruthless. On one particular day during my final year at the school, Brother Leo brought me into an empty classroom and shut the door behind him. From what I had been told, finding yourself alone with a Christian Brother was often a prelude to unpleasantness.

Fortunately for me, it was a mild agenda that afternoon. In fact, Brother Leo was looking for new recruits, and he asked me if I would consider becoming a Brother. At that time, my plans for the future included women, sex, lots of money, traveling the world, and maybe becoming an astronaut. And if I understood the question correctly, Leo was asking me if instead of those things, I might prefer to be locked up in a monastery with a bunch of strange men in robes.

I tried to think of a polite way to decline his invitation. It occurred to me that one of the requirements for joining the group might be a strong belief in such things as Jesus being the Son of God, or belief in God himself for that matter, along with miracles and everlasting life. So I respectfully let Brother Leo know that my faith in such things was perhaps not what it ought to be, but thanked him for his consideration. Brother Leo, however, remained unperturbed, and pointed out that many of the faithful had difficulty with their faith, and that this was nothing unusual – especially for an intelligent boy such as I. Faith was something that didn't just happen. It needed constant work and attention.

From what I could tell, matters of faith were all about accepting a lot of hearsay regarding purported happenings in the distant past. Taking the bible seriously meant that there really was a God – an omnipotent being that made everything happen. I didn't like believing in things. As a kid, I was far more impressed with methods of science. In fairness, having not yet conducted many scientific experiments of my own, I suppose I was still relying on second hand accounts regarding scientific veracity.

Yet the presumptions of science were usually based on fairly straightforward experiments and observations that ordinary people had made – observations that we could make ourselves, if we were inclined to do so. There was a common sense aspect to science; in many respects telling us what we already had come to appreciate about the repetitive and reliable workings of nature. If you fell out of a tree, it always resulted in an impact of some kind, along with some degree of discomfort.

In our religion class, we had a discussion about how God had created the universe. The aspiring scientists among us felt that it was more likely that the universe had always existed, since it was difficult to imagine a scenario in which everything was created all at once. What would have existed before God made everything? And what did God do for the infinite span of time before the universe was created? Did he, in fact, create time as well? We were skeptical of the Genesis account of how God first created light, and then everything else. Surely he would have first created stars - otherwise where would the light have come from?

Strange as it may seem, when I learned more about what science had to say regarding the origin of the universe, it was a closer match in some respects to the Genesis account than to our childhood, common sense viewpoint. Cosmology, it seems, is one area of science in which common sense often gets thrown out the window. Observations in fact suggest that the universe hasn't always existed; at least not in its current form. Instead it came into being some 13.7 billion years ago, with an event called the Big Bang.

So, in a sense, there was a creation event, and as if in deference to the divine decree "Let there be light," the early universe is, in fact, filled with light – or more precisely, high energy photons. It is a fair point that while science can help us to understand the way things are, religion attempts to do far more on the question of why. We may, for example, develop a quantum mechanical explanation of gravity, but still get no closer to satisfying the human desire for an underlying purpose to the physical laws that govern the universe.

Brother Leo argued that God's love can be directly felt, and is as real and tangible as the forces of nature. But that still requires that our feelings be interpreted through the lens of initial assumptions concerning the existence of a God who is responsible for creation, and is consumed with the affairs of humankind. That part really does require a leap of faith.

Faith and belief are words which have never much impressed me. I've always been more comfortable with words such as knowledge and hypothesis. But it would be disingenuous to suggest that one can live without faith. If there are no gods, then we must inhabit a cold and uncaring universe, completely at the mercy of the random happenings of nature. It's hard to swallow a universe like that, even for agnostics.

In the movie based on Carl Sagan's fictional work, *Contact*, astronomer Ellie Arroway submits that if humans are the only communicative beings in the cosmos, it "seems like an awful waste of space."[1] Either way, the universe is by and large a hostile place, filled with burning hot stars and barren irradiated planets interspersed through the vast cosmic emptiness. If you were dropped at random into the universe, chances are you would quickly perish.

It is sobering to think of the fragility of the human body. Each of us is but a heartbeat away from oblivion. As science writer Lisa Melton observes, "The dose of anesthetic that puts you to sleep is not much smaller than the dose that can kill you."[2] In a world without gods, there are no angels to flock to your assistance, should you suddenly suffer a coronary. On the other hand, if you were to collapse in a populated area, it's possible that help would arrive, maybe in the form of an ambulance full of paramedics who would jump out and try to revive you.

In other words, the world may not be such an uncaring place, as long as we have a little faith in our fellow humans. Secular humanists aside, there is often a sad lack of faith in what people can accomplish. It's even been suggested that aliens must have been involved in constructing the pyramids, or erecting the huge stones on Easter Island; the assumption being that people were too stupid or inadequate to have managed these feats without extraterrestrial help.

The reality is that unless there are gods, the only hope for humanity is humanity itself. As such, our hopes for salvation and for an end to human suffering depend upon the survival of our civilization. What is the likelihood that the human enterprise will survive? That's a difficult question to answer, since we do not, as yet, have any other sentient civilizations with which to compare ourselves.

If someday, we manage to detect an alien signal, then we could justifiably infer that the extraterrestrial technology is at least as advanced as ours, since they are obviously capable of interstellar communication. But given the range of possibilities, since they could be thousands, millions, or even hundreds of millions of years ahead of us, it's likely that such a signal will have come from a society that is much older than ours. For this reason, any positive result from the Search for Extraterrestrial Intelligence (SETI) would provide a tremendous uplift for humanity, regardless of the content of ET's message. We would have unequivocal proof that a society such as ours can have a future.

In 1951, SETI pioneers Frank Drake and Carl Sagan commandeered the giant radio telescope at Arecibo, Puerto Rico, to listen for alien signals from several nearby galaxies. By pointing the telescope at galaxies, rather than individual stars, they could cover hundreds of billions of stars at once. Even if only a small fraction of these stars had inhabited planets, they could be listening in on tens of millions of communicative civilizations. Surely one of them would be sending a signal in our direction. The astronomers listened patiently for days, but heard nothing from the aliens.

It was an outcome that Sagan found downright depressing, and as science writer William Poundstone recalls, it was enough to cause Soviet astronomer Iosif Shklovskii to question the whole premise of SETI.[3]

The negative results from Arecibo can be interpreted in a number of ways. Even if we suppose that there are no communicative aliens out there at this time, that doesn't mean that they never existed. Rather it could indicate that civilizations such as ours are relatively short lived, so that the probability of finding one at any given time is pretty small. This is not exactly a comforting interpretation, since it doesn't bode well for our species and our chances for long term survival.

It may also be a little premature to jump to any conclusions just yet, given that our investment in SETI has been fairly limited to date. For one thing, we do not as yet have the technology to eavesdrop on alien communications. Broadcast television signals have a tendency to leak into outer space, where they continue to travel, unimpeded through the distant reaches of the galaxy. This means, for example, that the first broadcasts of *I Love Lucy* from 1951 are just now beginning to reach our nearest stellar neighbors. No doubt the inhabitants of distant worlds are at this very minute struggling to understand the antics of Lucy and Ricky.

The reverse is also true – namely it is possible that at this very minute, ET's version of *I Love Lucy* is streaming past the Earth. Unfortunately, our radio astronomers are not equipped to detect alien television. But before we attempt to remedy this by covering the Sahara with television antennas, we should probably ask ourselves if the aliens are still using transmitter towers, or if, like me, they have switched to cable.[i][4] The only signals that Drake and Sagan would have been able to detect were ones that were either extremely powerful by our standards, or were beamed directly at us. Perhaps the most optimistic assumption that goes along with SETI is that aliens are deliberately trying to contact us using radio waves. If that's not what advanced civilizations do, then SETI is a bad investment for now.

There's no question that hearing from ET would be a momentous occurrence. Knowledge obtained from a benevolent race could perhaps put us on the fast path to discovering new cures for disease, or maybe help us to sort out some of our social and political problems. In the most hopeful portrayal of alien encounters, science fiction movies such as *Cocoon* suggest that we might even learn the secrets of immortality – a futuristic take on the fountain of youth theme.[5]

In truth, odds are pretty slim that aliens will come to our rescue any time soon. If our cosmic friends have a non-interference directive, like the one employed by the fictitious *Star Trek* federation, then we won't be hearing from them until we develop warp drive. There is, of course, the bleak possibility that we are alone in the galaxy. In all of our astronomical observations to date, there are no hints of spacefaring

---

[i] Astronomer Seth Shostak estimates that, at a distance of fifty light years, you'd need to cover an area of three thousand acres with television antennas just to pick up a TV signal. To actually interpret the signal, the antenna array would have to be "tens of thousands of times larger."

aliens; no fleets of robot spaceships, no vast extraterrestrial engineering projects, and no giant obelisks on the surface of the Moon.

On the other hand, our ability to see such things is fairly limited at the moment. Although we are beginning to detect planets beyond our solar system, we are not yet able to image them directly, although that is likely to change in the near future. Within the next few decades we'll have telescopes that will not only provide images of extra-solar planets, but they will also tell us which of these distant worlds harbors life. Spectroscopic detection of atmospheric ozone would be a telltale clue. The discovery of life beyond Earth will undoubtedly boost the confidence of the SETI folks, and we'll feel a little less lonely as a result. However, the search for life in the cosmos may end up being a much safer bet than the search for intelligence. Evolutionary biologist Ernst Mayr argued that, while the universe may be rippling with life, civilizations like ours could be extremely rare.[6]

Earth's early history was a tumultuous time. The planet was pummeled so intensely with meteoroids that the surface remained hot and molten. This hailstorm of rocks lasted for a billion years, until about 3.8 billion years ago when conditions improved, and things finally began to cool down. Once the heavy bombardment lifted, fossilized evidence suggests that life took hold at an incredible pace.[7]

Primordial Earth would have had an atmosphere composed primarily of nitrogen and carbon dioxide. Like the present day Saturnian moon Titan, there would have been a ready supply of the basic chemical constituents of life. One key ingredient that Titan doesn't have is liquid water, essential to facilitate the molecular interactions necessary for life. Water provides the stage, or platform upon which life's molecular actors perform.

Within this fluid environment, molecules can socialize and get to know one another, and they can also team up to form larger, more complex molecules. Amazingly, the process of life is mostly about the shapes of molecules; like tiny LEGO constructions, they only fit together when their shapes match up. In fact, drug discovery is often about trying to find an agent with just the right shape to fit into a specific type of cellular receptor, like a lock and key - thus blocking, or enhancing the action of some biochemical process.

Keeping this molecular dance in motion requires energy, which today's organisms can get from heat, sunlight, or by eating food and releasing the chemical energy contained within. Sulfides provide an important source of energy for organisms that live near hydrothermal vents on the ocean floor. Some of these deep sea dwellers are anaerobic, which means that they can survive without oxygen. In that sense, they bear some similarity to their ancient oxygen deprived forerunners.

Figure 3. Arecibo Observatory, Puerto Rico. Courtesy of the NAIC - Arecibo Observatory, a facility of the NSF.[8]

Another requirement for biology is catalysis. A catalyst is a molecular entity that can act as a matchmaker of sorts, bringing components into close proximity, so that they can react and make magic together. Without these catalysts, or enzymes as they are called, the molecules find it much harder to connect, and the chemistry of life is exasperatingly slow. Nobel laureate Renato Dulbecco notes that in the absence of enzymes, "It must have taken thousands of years to accomplish what a living cell can do today in a matter of seconds or minutes."[9]

One can imagine a simple scenario in which an enzyme acts as a sort of double-ended wrench; able to clasp two reactants, and hold them in place at the correct orientation so that they can snap together. Before such enzymes existed, the earliest biochemical factories may have been established within layers of clays, or on the crystalline surfaces of rocks that would have served as catalysts or scaffolds for molecular assembly. This may also help to explain how elements from rock minerals ended up being incorporated into biology.

Perhaps the most recognizable feature of life is its ability to self-replicate, a function that involves the copying of genetic information stored within a double stranded structure known as DNA. In the initial stages of life's development, before the evolution of DNA, another simpler structure known as RNA may have had an important role to play. RNA also has the ability to carry genetic information and to self-replicate. But RNA is a single stranded molecule that can fold back onto itself to form all kinds of interesting shapes, which means that in addition to its self-replicating ability, it can also act as an enzyme. These useful qualities could have allowed an RNA-type molecule to serve as a stepping stone on the path to DNA.

Some molecules show an intense dislike for water, and are called hydrophobic. Hydrophilic molecules on the other hand are very fond of water. Most curious are those that exhibit a sort of dual personality; hydrophobic on one side and hydrophilic on the other. A double layer sandwich of such molecules has a natural tendency to arrange itself into little bags of water, or globules, and it's not hard to imagine these structures as the precursors of living cells - another clue as to how life might have arisen from lifeless chemistry.

In 1952, in an effort to shed more light on the origin of life, Harold Urey and Stanley Miller performed a now famous experiment in which they used a sealed laboratory apparatus to create a miniature version of the primordial Earth. A flask was partially filled with water to represent the ocean, and this was connected via a tube to an upper vessel in which the supposed constituents of the early atmosphere were introduced.

Heat was applied to the "ocean" to produce water vapor, thus hopefully reproducing atmospheric conditions that would have existed some four billion years ago. The experimenters used sparks of electricity to simulate ancient lightning storms and then watched to see what would happen to the mixture. The water quickly turned yellow, and after a few days, the walls of the upper vessel were coated with a dark organic tar. It probably looked something like the fountain in my back yard.

Despite the simplicity of the experiment, Urey and Miller had cooked up a soup of organic molecules, including amino acids (which, like Bill Bryson, "I am obliged by long tradition to refer to here as 'the building blocks of life'").[10]

While nobody has yet managed to repeat such an experiment and have something crawl out of the test tube, almost all of the basic elements of living cells can, in fact, be created in the laboratory.[11] Scientists have even managed to produce synthetic DNA. However, with the benefit of hindsight, we can now say that Urey and Miller were probably wrong about contents of the initial atmosphere. It also seems unlikely that life's more complex molecules could have been sparked into existence by lightning storms. Most biochemists now have a more nuanced view of how such macromolecules arose, but there many more dots to connect before we can give a comprehensive account of how life's particular brand of chemistry can materialize from non-living ingredients.

Indeed, given the extent of molecular complexity over the past four billion years, there is still ample room for speculation on the exact sequence of steps that led to the emergence of our earliest ancestors.

In any case, the stuff of life appears to be abundant in the universe. In 2009, NASA announced that the amino acid glycine was discovered in a cometary sample returned by the Stardust space probe. Organic molecules have even been found drifting in the depths of interstellar space. The conditions that allowed life to flourish on Earth might therefore have been replicated on a multitude of worlds across the galaxy.

One might suppose, given the speed at which life got started on our planet, that there may be few obstacles to the emergence of life, once conditions are right. As quickly as the real estate becomes available, life is ready to move in. For these reasons, Mayr concedes that "it would seem quite conceivable that life could originate elsewhere in the universe."[12]

While it may not take a leap of faith to imagine a galaxy that is teeming with life, intelligence, however, is a different matter. As NASA scientist Chris McKay likes to point out, microbes are the real success story of evolution. The animals and plants with which we are most familiar, did not show their faces until relatively recently – within the last half billion years. Before that, the development of complex life was hampered by the lack of oxygen in the atmosphere.

We owe our existence to the efforts of cyanobacteria, working diligently for two billion years to get the atmosphere ready for our respiratory lifestyle. Yet, even when more complex animals appeared on the scene, nature showed little urgency in getting on with the job of making intelligent beings. If we adopt SETI's definition of intelligence, meaning the ability to manufacture radio telescopes or optical beacons, then the entire 150 million year reign of the dinosaurs, for example, represented zero progress toward this goal. It was a complete waste of time.

One might wonder if nature was ever serious about meeting the SETI target, or if the appearance of human civilization some ten thousand years ago was a surprise. Throughout the four billion years of its existence, our planet has been dismally quiet in the radio spectrum. Surely, says Mayr, we should expect the same silence from other life bearing planets.

Nevertheless, it would be the height of arrogance to suggest that humans are the only intelligent species to have evolved in the galaxy, which brings us back to the question of longevity. How long do civilizations typically last? As a westerner, when I ponder a question like that, and imagine what the future may bring, my projections are usually tied to the idea of progress. Walt Disney's "Carousel of Progress" was, for me, the most engaging representation of this idea that technological advancement is the key to a better future for everyone. Its theme song still readily comes to mind:

> There's a great, big, beautiful tomorrow,
> Shining at the end of every day.
> There's a great, big, beautiful tomorrow,
> Just a dream away.[13]

In post war times, all the energy that was put into making tanks, airplanes and ships was redirected into making gadgets and gizmos for the masses, and consumerism came into its own. Life has surely improved as a result. My grandmother, for example, spent a ridiculous amount of time washing clothes and scrubbing floors. She used fairly low tech devices such as scrubbing boards, and mangles. When I inquired as to what a mangle was, I learned about how Granny got her hand caught in one once, and was sorry I'd asked.

As a child, it seemed to me that no one doubted the promise of technology and free enterprise. The usual question directed at me was "What are you going to be when you grow up?" I didn't know what I was going to be, but the expectation was that I would produce something, or achieve something that would better the human race, and get the family some well-deserved recognition and fame. The challenge for us kids was to make something of ourselves, and not end up as slackers. Somehow my mother was always well apprised of the entrepreneurial goings-on in the neighborhood. As she folded the laundry she would casually mention how Bobby Gillis from across the street had established a nice business for himself, installing mobile phones in cars. And why couldn't I do something like that? If her intention was to spur me into action, it always had the opposite effect, and made me despise Gillis even more.

My dad, Jim, was the perfect example of a self-made man. He was born in 1944 in a house on Kildare Road, and spent his early years in a tenement on St. Stephen's Street. His family was not rich, and his childhood, to my mind at least, seems to have been fairly tough. He often mentions one particularly vivid memory of being locked in a trunk by his mother, and left there for hours.

At the age of ten, he built himself a small wooden cart, and then inquired door-to-door to see if any elderly neighbors wanted their fuel delivered. He got a shilling for each cart load of turf logs that he dragged back from Windmill Road. When he was twelve, he left school to look for work, and with characteristic determination, he stood outside the gates of Dublin Dairies every morning at 6:00am, calling out to the delivery trucks as they trundled by, until finally one morning, someone gave him a job.

Although he was the runt of the litter, Jim seemed to have more than his share of spunk and determination. When his older brother was being beaten up by a crowd of bullies, the little guy came running to the rescue, dived into the melee and fought them off with his fists. His parents encouraged him to become a tailor, on the grounds that he was sure to make a decent living, since people would always need clothes.

So, following their advice, he signed on with Polykoff's as an apprentice cutter. They manufactured gents' jackets. First the various components were cut from layers of fabric, and then these were sewn together and shaped. Dad's job was to shape the jackets, and he was expected to complete forty eight per day, the number having been decided by the union.

After a time, he found that he could easily manage sixty or seventy per day, but the union representative refused to let him enter them in the book. Before too long, my father had finished about an extra week's work that hadn't been logged, until finally he decided, despite the union's directive, to just log the extra work in the book anyway. He remembers being approached by Seamus O'Riordan, the company foreman, who asked if there had been some mistake.

"No, no mistake," was the reply, and Jim pointed out the work he had completed. They must have been impressed with my father's contribution, because they quickly decided to promote him to the position of trainee foreman. However, when Jim learned that the union intended to block his promotion because, according to their protocols, the promotion should have gone to the next person in line, he promptly resigned in disgust. If that was the way unions behaved, where was the motivation for anyone to show some initiative?

Dad believed that the way to improve one's lot in life was through hard work and individual determination. So over the years, as he moved into management, my father did pretty well for himself. He supported a brood of five kids, was the first in his family to become a home owner, and eventually managed to buy a vacation home on a few acres of land near the coast.

"Not bad," he used to say, "for someone who grew up in a tenement on Stephen's Street."

His personal philosophy was straightforward.

"Do the best you can for yourself and your family," he told me, "without trampling on anyone else in the process."

I remember asking him if capitalism was unfair, since it meant that some people would always be far wealthier than others.

"Well if you work harder, then you deserve to be rewarded," was his reply.

As far as the poor were concerned, Dad explained that everyone needed to have an equal chance, and that was why we needed government to provide decent education and healthcare; so that we had a level playing field for everyone. He didn't believe in unemployment.

"There's always work to be done," was his motto.

This was my introduction to how the world worked. It was obvious that what every country needed was a hearty helping of democracy and free enterprise, along with a laissez-faire economic policy. This would allow people to make their way in life and be successful. While some people squabbled over inheritance rights, others could do just as well if not better for themselves, having started out with nothing. When politicians tried to take the credit for more prosperous times, Jim was cynical. In his view, when businesses succeeded, it was the work of managers and investors, not politicians.

"The only thing the government can do is to create a good environment for business, preferably by keeping the tax burden low."

These days, there are few who would seriously question the ethos of capitalism. My mother tells me that in Granddad Waine's time, kids would go to school with no shoes on their feet. Milk was delivered door to door by horse and cart. Now, thanks to progress, kids not only have shoes; they also have cell phones and personal computers. The demise of the Soviet Union in 1991 was widely regarded as the final victory for the capitalist way of life – a feather in the cap of free enterprise.

The United States has become the bastion of capitalism – not so remarkable perhaps, given the historical influx of immigrants seeking religious and economic freedoms. The strength of the capitalist system lies in its simplicity. In the earliest years of the nation's history, there was no need for any master plan for how people should live, with everything figured out to the last detail. Instead, success depended solely upon ingenuity and entrepreneurial spirit, both of which could be found in abundance.

America was seen as a place of unlimited potential – a magnet for those with ambition and drive. In the second half of the nineteenth century, the US displaced Britain as the leading industrialized nation in the world. Productivity soared, as did the number of inventions. The Gilded Age, as it was called, was a manifestation of America's love affair with business, and it was a time in which commercial interests came to dominate every aspect of our politics and culture.

There can never have been a time when technology seemed more wondrous. With the completion of the transcontinental railroad in 1869, you could travel from coast to coast in just 11 days. In the preceding decades, the same trip would have taken months by covered wagon. The magic of the telegraph allowed people to communicate almost instantaneously over long distances, and with the later discovery of radio, it was even possible to talk with ships at sea. Industry made its presence felt through a vast expansion of the country's transportation and communications infrastructure, while natural resources were extracted at an astonishing rate. The US produced more steel than Britain, Germany and France combined, and coal consumption was up eight hundred percent.

A less obvious, but equally important aspect of industrial expansion was the adoption of the corporation as a vehicle for private industry. Corporations had existed previously, but had generally been used by governments for large public works projects, such as the construction of new dams, turnpikes or aqueducts. Businesses, on the other hand, were run by their owners, which limited their opportunity for growth. This dependency meant that if, for example, the owners had a falling out, or if one of them died, business could be badly disrupted.

The decision to allow corporations to be employed for private entrepreneurship removed many of these constraints on growth, and allowed businesses to develop unhindered by human limitations. In the 1800s, this was such an attractive proposition to budding industrialists that states were increasingly pressured to legitimize the use of corporations as agents for personal gain. In 1886, the US Supreme Court ruled that corporations should have the same inalienable rights and privileges as individuals, and a new species was set free to stomp the earth.[14]

There was a sense of foreboding among some of the onlookers. Corporations are strange individuals, and they bear some distinct advantages over humans. For one thing, they can have inordinately long lifespans – some may live indefinitely. They tend to be more focused than the rest of us, and have a single minded outlook on life; namely an obsession with turning a profit. Perhaps the most worrisome distinction is that a corporation is not burdened with a conscience, so it never gets distracted by ethical ruminations. This was a concern even in the nineteenth century, when no one could have imagined how this new breed of individual would come to dominate the geopolitical landscape.

> Corporations will do what individuals would not dare to do.[15]
> - Peter C. Brooks, 1845.

The electrification of industry brought mass production techniques to full throttle, and made way for the steady procession of technological marvels that continues to this day. As new competencies were applied to every field of human endeavor, there could be little doubt that capitalism was working, especially in the eyes of leading industrialists who became extraordinarily wealthy.

However, the Second Industrial Revolution was also a time of great poverty, as the gap between rich and poor widened into a chasm. The new American aristocracy adopted a dog-eat-dog brand of free enterprise that allowed the poor to go to the wall for the greater good of progress. In the eighteenth century, manufacturing was made more efficient through the introduction of factories, which also provided new employment opportunities for city dwellers. From the outset, however, there was little regard for the people working in these factories. Working hours were long, and conditions unhealthy.

When workers pushed for a ten-hour work day, business owners vehemently opposed the idea, arguing that it would drive men to mischief to have so much free time on their hands. The arrival of machines into the work place allowed factories to operate around the clock, making conditions even more hazardous for the workers, and leading to high numbers of casualties. Historian Alan Brinkley writes that "children working at the looms all night were kept awake by having cold water thrown in their faces."[16] "Whipping rooms" were provided for disciplinary purposes. Rhode Island scholar Maury Klein notes that "Although accurate data on workplace accidents do not exist before

1920, one estimate claims that an average of 35,000 workers were killed and 536,000 injured every year between 1880 and 1900."[17] A newspaper article from 1911 reports that "in one factory one summer seventeen men contracted typhoid fever because of insanitary conditions."[18] The New York State Factory Commission interviewed a woman who, in 1909, barely managed to live on $6 a week.

> I didn't live, I simply existed. I couldn't live that you could call living. I certainly had to deprive myself of lots of things I should have had. It took me months and months to save up money to buy a dress or a suit or a pair of shoes.[19]

Employers provided workers with accommodation, as long as the entire family worked in the factory. Children were forced to work under horrendous conditions. A report submitted to a factory inspectors' convention in 1895 gives a particularly gruesome example.

> What could be more revolting than the presence of three hundred children in the Chicago stockyards, scores of them standing ankledeep in blood and refuse, as they do the work of butchers?[20]

Appalling as such observations were, photographs proved to be more effective at pricking the American conscience, in particular those of Lewis Hine, who worked with the National Child Labor Committee to help spur efforts to end the exploitation of children in the workplace.

This type of extreme capitalism was lauded by such corporate greats as John D. Rockefeller, who believed that his ruthless business practices were simply a reflection of how nature worked, and thus were endorsed by God. "God gave me my money," he said.[21] Englishman Herbert Spencer, who coined the term "survival of the fittest," held that the principles of Darwinian selection were equally applicable in matters of sociology and economics. Just as organisms compete for survival, businesses compete with each other for market share, and customer loyalty. Those that survive are, obviously, the "fittest."

By giving individuals the freedom to conduct business without interference, we were, in effect, using the same laws employed by nature over the millions of years. Spencer opposed attempts by government to regulate industry, because they upset these natural laws. In his view, the poor should be allowed to perish.

> Beings thus imperfect are Nature's failures, and are recalled by her when found to be such. . . . If they are sufficiently complete to live, they *do* live, and it is well they should live. If they are not sufficiently complete to live, they die, and it is best they should die.[22]

After all, as Baptist minister Russell Conwell often proclaimed:

> Let us remember there is not a poor person in the United States who was not made poor by his own shortcomings.[23]

If the process of natural selection had led to wondrous developments in biology, didn't it also make sense to use nature's methods in plotting the course of civilization? By drawing on Darwin's theory of Natural Selection, proponents of unrestrained capitalism could more easily make the case that their ideas were grounded in science. It was a point of view supported by Yale professor William Sumner, who wrote:

> We have already become familiar, in biology, with the transcendent importance of the fact that life on earth must be maintained by a struggle against nature, and also by a competition with other forms of life. In the latter fact biology and sociology touch.[24]

"The millionaires," said Sumner, "are a product of natural selection. . . . In this respect they are just like the great statesmen, or scientific men, or military men."[25]

On the face of it, there seem to be real similarities between free enterprise and natural selection. If two bookstores open up in a town where there is only enough business to support one of them, which will survive? Presumably, the store that stays in business had some advantage over the other, even if the advantage was small. Perhaps it was better situated to attract customers, or maybe the proprietor was able to build a better business relationship with the local clientele.

Let's imagine a scenario in which one of the owners, Doug, is doing a little better than his competitor, Tony, until, after a freakishly heavy storm, the local river overflows, putting Doug's store under three meters of water. With the subsequent disruption to business, Doug finds himself at a disadvantage, and he loses so much business to Tony that eventually he is forced to close. Doug had a better head for business, and was well liked in the town. If it hadn't been for the flood, Tony's store might have been the one to fail. The fact that it was situated at a slightly higher elevation than Doug's, might never have mattered, but when circumstances changed, this otherwise irrelevant feature gave him the advantage.

Figure 4. Photo taken by Lewis Hine in 1909, shows mill children in Macon, Georgia. "Some boys were so small they had to climb up on the spinning frame to mend the broken threads and put back the empty bobbins."[26]

Likewise, from an evolutionary perspective, it can be difficult to predict who will survive in the long term. It often happens that when things change, some fortunate species finds itself, quite by accident, to be particularly well adapted to cope with the new circumstances. The demise of the dinosaurs provides a dramatic example. These creatures were superbly adapted to their surroundings, and many types of dinosaurs had been roaming the planet for 150 million years. Their large size gave them an advantage over other species such as mammals. It is now thought that climate change, due to the impact of a large asteroid, may have made life very difficult for the dinosaurs, thus explaining their rather sudden disappearance sixty five million years ago.

The discovery of a 180 kilometer wide impact crater under the Yucatán Peninsula in Mexico, suggests the asteroid was about ten kilometers in diameter. Such an impact would likely have blasted dust and debris into the atmosphere, shrouding the planet and preventing sunlight from getting through. The result would have been a prolonged, severe drop in surface brightness, causing plants to die. Large animals would have been hardest hit by the food shortages, whereas small, warm blooded mammals were now better adapted for survival.

These examples illustrate some important differences between the ideas of laissez-faire economics and the actual workings of natural selection in biology. Over the course of evolution, the degree to which a species adapts to its surroundings has nothing whatever to do with what individuals might want, or decide. A successful species is one in which random chance has operated favorably, for the time being. Although the process of natural selection is often described as a struggle for survival, the potential of a species, and indeed its very existence, depend upon chance events – not upon how much it struggles. It has very little to do with ingenuity or hard work.

In the world of nature, the struggle for life is opportunistic, but without purpose or rationality. The world of business, on the other hand, is a world of our own making; a world in which people can make choices. As President John F. Kennedy famously said, "Our problems are manmade; therefore they can be solved by man."[27]

Capitalism, by its very nature, is a system that places freedom above equality, so surviving in business will always amount to a struggle, particularly in difficult economic times. In 2002, the PBS show *Now* told the story of a young woman who was trying to support a family while working as a waitress. When she lost her job, she decided to train to become a nurse, but then discovered that she couldn't collect unemployment benefits unless she gave up school.[28] As rational decision makers, we can use our legislative framework to alleviate some of the stresses caused by changes to the economy. We could choose, for example, to establish a policy that helps people to acquire the training and skills they need to adapt to new business conditions.

Humans are now changing the environment so quickly that we have outpaced biological evolution. Our chances for survival are thus far more dependent upon the pace of technological innovation and the rapidity with which the world changes as a result. If, as Sumner suggests, human activity can be characterized as a "struggle against nature," should we then rely upon natural laws of random chance to reshape our society?

Sumner argues that any attempt to interfere with the evolution of society will only result in "the survival of the unfittest."[29] But can we be sure that what's good for the Rockefellers and Vanderbilts will also be good for the long term survival of humanity? For that matter, can we be sure that the wealthiest people are in fact the fittest members of society, and that the poor are inherently unfit? It's not simple, say sociologists Laura Desfor Edles and Scott Appelrouth, since "both 'success' and 'failure' hinge not only on individual aptitude and effort, but also on institutional and cultural dynamics that sustain a less than level playing field."[30]

In the sense that we can use the institutions of government to help even the score for everyone, it would seem that in the modern era, the fitness of civilization increasingly depends upon sensible public policy making. On the other hand, poor public policy can have long reaching and sometimes devastating consequences. A well intentioned energy policy decision, for example, to allow the reprocessing of spent fuel from nuclear power plants could make it easier for terrorists to acquire separated plutonium.

As far as evolution is concerned, the development of language is believed to have played an important role in the emergence of symbolic thought and creativity, and the unique way in which humans think about the world – surely the source of that individual determination that Social Darwinists were so earnest about.[31] It is ironic that these attributes should themselves be a product of language – a social composition. Without a high degree of collaboration in cohesive social groups, individualism would never have evolved in the first place.

At the dawn of the twentieth century, there were few restraints on the proliferation of industry. Up to this point there had been very little involvement by government, and businessmen were free to do battle as they saw fit. This made it easier for businesses to expand, often by savagely eliminating whatever obstacles they encountered. Rockefeller built his empire by removing competition. He established a trust that facilitated secret collusions to hold prices at artificially high levels, while he benefited from a crooked system of rebates and kickbacks. Those that did not fall into line were driven out of business.

> The growth of a large business is merely a survival of the fittest. . . . This is not an evil tendency in business. It is merely the working out of a law of nature and a law of God.[32]
> - John D. Rockefeller, Jr.

Klein tells us that Rockefeller became so powerful that he was able to "[bully] the railroads into giving him rebates on the shipment of his own oil *and* that of rival companies as well."[33] The tycoons of this era amassed unprecedented levels of wealth and influence. Although they usually attributed their good fortune to hard work and determination, perhaps a larger share of the credit should be given to ruthlessness and arrogance, along with a willingness to engage in bribery and other cutthroat tactics.

There was little or no concern for the effect that their corrupt business practices were having on the American public. Farmers, for example, became increasingly distraught at being "railroaded" out of existence by the exorbitant prices they were being charged to transport their produce. But the new American aristocracy was answerable to no one. Their attitude towards the rest of America was nicely encapsulated in William Vanderbilt's famous remark, "The public be damned." When his father, Cornelius was betrayed by two of his associates, he is reported to have told them:

> Gentlemen: You have undertaken to cheat me. I won't sue you, for the law is too slow. I'll ruin you.[34]

These abuses became so extreme that the resulting public outcry forced the government to play a bigger role in regulating business. The passing of the Interstate Commerce Act, in 1887, and the Sherman Anti-Trust Act in 1890 marked a change in attitude toward the role of government. During the Progressive Era of the early 1900s, public policy was increasingly used to rein in the excesses of corporate greed in the interest of the common good.

Higher wages and more generous benefits contributed to the emergence of an affluent middle class. With more money to spend, these families satisfied their appetites for new life enhancing products, which of course led to increased demand, thus inflating the cycle of consumerism. Bryson describes how 1950s families couldn't help but betray their new found pleasure and infatuation with new household appliances.

> When I was about four my parents bought an Amana Stor-Mor refrigerator and for at least six months it was like an honored guest in our kitchen. . . . When visitors dropped by unexpectedly, my father would say: "Oh, Mary, is there any iced tea in the Amana?" Then to the guests he'd add significantly: "There usually is. It's a Stor-Mor."
> "Oh, a Stor-Mor," the male visitor would say and elevate his eyebrows in the manner of someone who appreciates quality cooling. "We thought about getting a Stor-Mor ourselves, but in the end we went for a Philco Shur-Kool. Alice loved the EZ-Glide vegetable drawer and you can get a full quart of ice cream in the freezer box. *That* was a big selling point for Wendell Junior; as you can imagine!"
> They'd all have a good laugh at that and then sit around drinking iced tea and talking appliances for an hour or so. No human beings had ever been quite this happy before.[35]

By the end of the century, although more than a third of the country's capital remained in the hands of the wealthiest one percent, the standard of living had been raised for the rest of us as well.[36] Working conditions in all of the industrialized nations had improved dramatically. It was a century in which America had won two world wars, a space race and the fight against communism. Diseases had been eradicated, and life expectancy was significantly higher due to advances in medical science and public health. Works of the great performers were preserved for the first time in audio and video recordings. Injustices were uncovered and dispensed with and civil rights won, resulting in a world with more equality, more freedom of expression, and more opportunity for individuals to live out their potential.

All of this was captured, and brought to the world through the medium of television, providing citizens with the means to observe and adjudicate on the happenings of civilization from the comfort of their living rooms. It was an exhilarating ride for humanity. In less than a hundred years, we went from horse and buggy to space shuttle, a journey that included the development of nuclear power, and a visit to the Moon. The United States, starting out in pole position, emerged from the 1990s as the most powerful nation to have ever existed on the planet. Every field of human endeavor was taken to giddying heights. Scientific discoveries were nothing short of revolutionary. The stunning revelations of relativity and quantum mechanics gave us an entirely new perspective on the universe, and opened up a world of hitherto unknown possibilities in electronics and computing. Equally transforming discoveries in biology allowed scientists to unravel the secrets of DNA, providing us with the tools to understand the workings of nature, including our own biological underpinnings. For the developed world at least, as we savored the delights of accomplishment and prosperity, it was clear that the philosophy of progress had lived up to its promise.

While all of this progress was underway, my dad spent most of his years working in the rag trade, as it was called, which was one of the first areas of manufacturing to move overseas in search of cheaper labor. Improved logistics and lower shipping costs allowed other industries to follow suit. For business owners, this can be a smart move, as it can boost profits while still allowing products to be more competitively priced. But it can also create more problems in the developing world where, with less regulation and lower standards, working conditions may be no better than they were in nineteenth century America.

When the clothing industry went bust, Dad moved to another company that specialized in bathroom remodeling. On a trip to China, his boss, Pete, was given a tour of a factory where workers were spraying some kind of enamel finish onto bath tubs. He'd seen this being done before, during a visit to a plant in Germany, where most of the work was done by robots. In China, things were different. Here, workers were spraying the tubs by hand, and the air was filled with toxic aerosols. Pete was shocked that nobody was wearing any respiratory protection.

"Oh, don't worry," the tour guide assured him. "If they get sick, we can replace them very quickly."

This overseas relocation of manufacturing can create a race to the bottom, leading to environmental degradation and unhealthy conditions for workers in host countries. It can also lead to job losses in western nations. Tailoring jobs were among the first to disappear. My father often adds a few choice words whenever he reflects on his parents pushing him into tailoring. I remember asking him what we would do if all the manufacturing jobs went away.

"Well, in that case, we'll all have to become computer programmers," was his reply. "All that's left are the high tech jobs. Unfortunately," he mused, "this kind of work doesn't suit everyone."

Of course, it wasn't long before software development got shipped overseas as well. As it becomes increasingly difficult for middle class Americans to find good wages, spending power is reduced, resulting in even more pressure on the economy. While manufacturers look to new consumer markets in Asia, many environmentalists believe that our long term prospects look grim.

The underlying problem, they say, is with consumerism itself. We now refer to ourselves, not as citizens, but as consumers, chiefly because our economic growth is now inseparably tied to consumption. Increased consumption leads to diminished resources, and often results in the loss of environmental services – a loss that has not traditionally been factored into the cost of the goods we produce. For many, this is a recipe for doom.

In the rest of the world, we have a reputation for being somewhat wasteful. It was said in 1907 by Austin Bierbower that "two Frenchmen can live off what one American wastes and live better than the American."[37] Pulitzer Prize winner Edward Wilson warned that "For every person in the world to reach present U.S. levels of consumption with existing technology would require four more planet Earths."[38] Those of us, who not busy consuming, are relentlessly engaged in the frenetic drive to get products in front of consumers. The famous

American writer, Ralph Emerson, believed that such a fixation on commerce was unhealthy, and that it dulled the imagination.

> The mind of this country, taught to aim at low objects, eats upon itself. . . . Young men of the fairest promise, who begin life upon our shores, inflated by the mountain winds, shined upon by all the stars of God, find the earth below not in unison with these, - but are hindered from action by the disgust which the principles on which business is managed inspire.[39]

Yet we could hardly be more infatuated with trade. "The chief business of the American people," said Calvin Coolidge, "is business."[40] Citizens are called to serve their country, not through community service, but by going shopping. We Americans, it seems, live to work, rather than the other way around.

In the nineteenth century, even as bewildering levels of raw materials were extracted from the environment to satisfy a growing industrial appetite, it was inconceivable to most Americans that there might be limits to what nature could provide. Move the clock forward a century or two, and as the world starts to shrink, so does its repository of natural resources.

Perhaps the biggest factor in this new reality is population. For scientist Paul Ehrlich, the small African country of Kenya was a case in point. In 1968, the population was set to double within twenty three years, which meant that they would have to double their infrastructure; twice as many roads, hospitals, schools, and twice the amount of food and fresh water, "just to keep living standards at the . . . [same] inadequate level."[41] As Anne and Paul Ehrlich reported in 1990, Kenya's population did, in fact, double as expected.[42]

For some parts of the world, you can access sequences of satellite photographs showing the growth of populated areas over the decades. As I looked through some examples, the images reminded me of petri dishes in the laboratory, used to examine the growth of bacteria colonies. In particular, it made me think of the analogy presented by University of Colorado physicist Albert Bartlett, who imagined a container in which a population of bacteria doubled every minute. In his example, the experiment begins at 11:00am, and by noon, the dish is completely full of microorganisms. Interestingly, at 11:59, with just one minute left, only fifty percent of the real estate has been used up. At what point, asks Bartlett, would a single average bacterium look up and notice that space was running out? The answer is, probably never; or at least not until it was too late for the microbes to do anything about it.[43]

The growth of a population means more competition for the commodities of life, which in turn increases the pressure on society. Standards of living decrease, as do personal liberties. Freeways that once made us free now become clogged and dysfunctional with traffic. At a basic level, many of the problems that we are likely to face with a growing population, urban congestion, sanitation, disease, poverty, and drought, have been successfully dealt with by government programs in the past. Most cities, for example, have been able to provide basic sanitation and clean drinking water for their citizens. However, such public works efforts were often accomplished with general disregard for broader environmental effects, including the pollution of lakes and waterways, and the disruption of ecosystems.

As the pressure mounts to address global issues such as climate change and the overall shortage of fresh water, it would seem that there's almost a direct relationship between the intensifying complexities we face and the growing ineptitude of the traditional remedies we use to deal with them. Furthermore, the individuals who are best positioned to orchestrate effective public policy are not really individuals at all, but corporations whose sole motivation is profit.

Ours is a system whose most accomplished progeny are often strongly inclined to act in the interests of the public good – consider the charitable and philanthropic ventures of Bill Gates, for example. Yet, ironically, the most powerful entities in our society are those that represent the embodiment of greed and consumption. Theirs are the loudest voices, with the potential to drown out the needs and concerns of ordinary people.

Hopefully, a balance of sorts can be achieved. Just as we celebrate the checks and balances inherent in our system of government, perhaps our legislative framework will serve us well, supporting a battleground of competing corporate interests, eternally at loggerheads with a progressive citizenry. But what can we say about the notion of progress itself? Is it compatible with our long term survival?

To answer that question, it is worth considering the fact that not all civilizations have embraced our way of living. In *The Day the Universe Changed*, author James Burke draws a contrast between our western expectation of continuous improvement through technology, and the very different outlook of eastern civilizations that do not hold to this idea of constant change.[44] Naturally, we westerners are proud of our accomplishments, and we can boast about the many ways in which we've demonstrated our technical prowess – landing a man upon the Moon being a notable example of what we can achieve when we set our minds to something.

We cannot, as yet, boast supremacy as far as longevity is concerned. There are many non-human species on the planet that can lay claim to much longer lifespans. Dolphins, for example, have been swimming in the ocean in their modern form for at least four million years. During that time, their activities probably haven't changed much. In fact, with no major engineering feats to speak of, their list of accomplishments is disappointingly low for such a huge expanse of time.

Even in human terms, western civilization is a relatively recent phenomenon. The Australian Aboriginal culture has existed for at least 40,000 years, probably because of their close relationship with their natural surroundings. In the Aboriginal world view, continued existence was very much about preserving the environment – not changing it. Their language doesn't even have a word for time.[45] Traditionally, Aborigines used burning techniques to prevent the accumulation of dry brushwood which could lead to extremely hot and damaging fires; an approach which also helps to keep the land productive and boosts biological diversity. Researchers are now studying these land use practices in their attempts to manage resources more sustainably.[46]

As we continuously look to reinvent ourselves, the secret to longevity, it seems, is to make as little change as possible; to live in synchrony with the natural world, instead of struggling against it. Our civilization can be measured in hundreds of years - not tens of thousands. In that short slice of time, we have reduced Earth's biodiversity, squandered resources, overseen drastic changes to our surroundings, diverted the course of rivers, paved over fertile lands, introduced tens of thousands of novel chemicals into the environment, and flirted with nuclear energy.[47]

If longevity were a priority for us, then maybe we would play it safe, take a lesson from the Aborigines and avoid meddling with natural processes that have already stood the test of time. But instead, our western way of life is more akin to playing Russian roulette with the future. Certainly our appetite for change leads to a more exciting story, but it also invites instability, and puts us at greater risk in the long term. Ironically, perhaps Spencer and Sumner were onto something when they said we should model our activities after nature. But that's not the course we have charted for humanity.

A reasonable person might conclude that if we remain on the "Carousel of Progress," our chances for long term survival are poor. But let's take a more optimistic look. Ours is an unusual case; a unique scenario from nature's playbook. No communicative aliens have proffered their case, so we can't say for sure when the curtain will come down. It's true that the dinosaurs' strategy worked – they prolonged their collective lifespan by keeping their heads down, making sure not to destabilize the biosphere with dinosaur technology. But it's also true that they were wiped out by an asteroid. Our technology gives us the ability, not only to see these upcoming threats, but to do something about them. We have the means to detect an approaching asteroid while it is still in deep space, and to determine whether it is on an impact trajectory. If we were to spot one in a timely enough fashion, we might even divert it from its course and prevent the impact from occurring.

So, "other things being equal," as Carl Sagan would say, "it is better to be smart than to be stupid."[48] It may be that, when it matters most, our technology will save us. When Paul Ehrlich predicted in 1968 that the world would be hit by devastating famines, it turns out he was wrong. He had underestimated the potential of agricultural science. The number of undernourished people in the world actually dropped, even as its population grew.[49]

It's true that we haven't seen any examples of civilizations like ours on other planets, but that doesn't mean there aren't any out there. The search has just begun. And even if such civilizations are rare, it may be because it's hard for them to get started in the first place – not because they inevitably destroy themselves.

Granted that's a lot of wishful thinking, but in our quest for human salvation, maybe that's what's needed - a leap of faith. If there are no

Gods, or aliens to save us, then we must save ourselves, and the only way we will do this is through progress. We could choose to abandon our technology, and live a sedentary existence, but we would get no further in our quest than did the dinosaurs, or the countless humans who have already lived out their hunter-gatherer existence.

If we play by nature's rules, then we will surely die by them as well, just as did the tens of billions that came before us. That is not our way. Perhaps, among the multitude of stars, many civilizations have flickered into existence, with the same aspirations of grandeur, only to promptly extinguish themselves before reaching their potential. And it could be that progress will carry us along the same course, to blindly follow our celestial cousins to our eventual doom. But I prefer to imagine that we will survive, and that we will add something new to the story of existence; that our descendants will be the ones to answer the ultimate questions, and live the dreams of forgotten ancestors.

4

⌘

Being Human

In 1974, an astronaut was attempting to land an experimental aircraft when disaster struck. He was almost dead when they pulled him from the wreckage. Fortunately for him, government scientists came to rescue, and managed to rebuild him with improved bionic parts.

I'm referring, of course, to the hit TV show, *The Six Million Dollar Man*, which was aired in the 1970s. Steve Austin's new mechanical appendages looked incredibly realistic, complete with artificial skin, hairs, and tiny sensors that recreated the sense of touch, so that nobody could tell that he was less than human. In fact, as we learned from the voice of Richard Anderson in his role as Steve's boss Oscar Goldman, his bionics made him "better, stronger [and] faster" than he was before; able to leap to the top of a tall building, or to bend steel bars with ease.[1] His electronic eye allowed him to magnify distant objects, and to see in the dark.

This was one of my favorite shows. As is typical for kids, my friends and I considered ourselves to be more or less invincible – just like the bionic man. Nothing could harm us. I once played a trick on my younger sister, pretending to hurt my arm, to see if she might notice the wires and circuitry that now protruded through the sleeve of my sweater. The plan was to make her believe that I had a bionic arm, or that I was some kind of robot! As I remember it, her reaction wasn't as animated as I would have liked. I'm not sure if I managed to fool her, but either way she didn't seem all that impressed.

There were lots of wonderful science fiction shows in the seventies, providing us kids with a smorgasbord of ideas, and fueling our imaginations with boundless optimism for the future. I was a big fan of

TV in general, and a big believer in the promise of technology. Of course, in those days we also had a much higher tolerance for cheesy special effects.

It made sense that we were headed for a future in which we got a better deal than the one that was sold to our parents. Life was constantly changing and improving in ways that mattered. Our first remote controlled television was one such important change. You could now actually control everything from the couch, without having to walk over to the set. How cool was that? As I remember it, my sister and I calmly and politely took turns investigating the workings of the clicker. Then we were visited with more excitement when the first infrared version arrived. Now we didn't have to deal with that clunky mechanical sound, plus it had more of a space-age look and feel about it. We had at least three minutes of fun right out of the box, as we pointed the controller in different directions, and experimented with mirrors to see if the invisible beam could be reflected.

But the memory of those remote controls quickly pales as one remembers that glorious evening when Dad unexpectedly brought home a VCR under his arm. The future knew no limits. Now we would have entertainment on demand. The rest of the world's problems would certainly be fixed by the time we grew up. If you lost any limbs, it would be simple matter to get bionic replacements that were much better than the biological versions. They wouldn't break as easily, nor would they corrode or get infected. And if they did break, it would be easy to get spare parts and upgrades. It might even be possible to reduce the perception of pain in the event of an injury.

When it came to science fiction, my parents usually pretended not to notice what was happening in the plot, as it was unbecoming for adults to allow their imaginations to run riot. Occasionally though, you would hear a comment from the back of the room.

"I wouldn't mind having Dr. McCoy do something about this pain in my back."

There was very little that *Star Trek*'s Dr. McCoy couldn't cure. Whenever an alien pathogen was infecting the crew, "Bones" would usually have a remedy figured out before the fifty minutes were up. It was administered without the use of needles, using something called a hypospray, and it took effect immediately. Even if a body was completely covered in scars and lesions, one shot from the hypospray and every trace of the disease disappeared within a matter of seconds. As I got older, it came as a bit of a shock to find out that in reality, it could take ten years or more to develop a drug and bring it to market. New medicine very rarely provides a complete cure, and for some people, it doesn't work at all. Growing up on a science fiction diet means that as you accrue more years, you continue to be disappointed by the comparatively slow progress that's being made in the real world. Of course, the flip side is that you often come across scientists, artists and engineers who chose their particular careers because they were influenced by science fiction when they were young. Sometimes, fictional ideas may even serve as technical challenges, and can spur innovation. Props used on the original *Star Trek* show bear an uncanny resemblance to products that emerged in subsequent decades, such flip phones, medical scanners, tablet computers, optical discs, and Bluetooth ear pieces.

As compelling as such excursions from reality may be, these television shows are not really about making accurate predictions of the future. While there may be no limit to the imagination, science fiction is a product that must appeal to a contemporary audience, and must have an interesting, and dramatic story to tell. As such, there are limitations on the extent to which writers can keep it real. Consider, for example, *Star Trek's* transporter device; a matter-energy converter that's used to beam people from place to place. The concept was originally introduced as a way to avoid having to land the starship on whatever planets the space travelers encountered. Landing and take-off sequences would have cost more in terms of both special effects and running time, so it was easier to just have individuals disappear and then materialize a moment later on the planet's surface.

In the episode, "Relics," while exploring the wreckage of an old transport shuttle, the crew of the *Enterprise* discover that someone has managed to survive inside a transporter mechanism for decades. In *Star Trek* speak, the transporter has been "jury rigged" so that the "matter array" gets repeatedly passed "through a pattern buffer" in a continuous cycle.[2] Luckily the information in the transporter buffer hasn't degraded, and they are able to complete the materialization process and restore the trapped individual; converting his energy pattern him back into a living, breathing human. The character in question happens to be none other than Montgomery Scott, Chief Engineer from the 1960s *Star Trek* series, apparently unaffected by the seventy five years he spent in the machine.

It would appear then, that the transporter has even greater potential than one might have realized. This amazing device has the ability to capture the essence of a human being, with memories and personality intact, and store this information in some kind of memory buffer. In the normal course of events, the transporter buffer is used only briefly, while the payload is in transit. As fans will be aware, there are a number of episodes in which this feature serves as a useful theatrical tool for creating a little extra dramatic tension. Recall that in the original series episode "Mudd's Women," for example, the transporter is used to rescue passengers from a ship that's about to explode. Thanks to the buffering ability, we see the ship actually explode before the people begin to materialize on the platform; so there are a few nervous moments while we wait to see if Mr. Scott managed to retrieve everyone quickly enough.[3]

We've seen the transporter used in other circumstances to decontaminate crew members who are beaming up from unknown environments, and who might have been infected with alien microbes. This decontamination is achieved using previously stored patterns of the folks who are beaming up, to filter out any pathogens that might also be present in the incoming stream.

If the transporter can hold onto the patterns of individuals in transit, it would seem that the same technology could be adapted to create and store copies of these patterns. Why then isn't it standard practice to create back-up copies of Starfleet personnel before beaming them into hazardous situations? In this way, if the captain ends up getting eaten by a giant lizard, he can be resurrected from the latest backup. No doubt there are a variety of technobabble arguments that explain why this isn't supposed to be possible. But couldn't *Star Trek*'s twenty fourth century engineers figure this out if they put their minds to it? After all, this is the basic premise used for the episode, "Second Chances," in which we see two versions of Commander Riker.[4]

Ok, I admit to being a little facetious here. We know there are good reasons for not writing this sort of thing into the storyline. Removing the possibility of death is not always a good thing when you're trying to keep your viewers on the edge of their seats. It would be hard to shed a tear for Captain Spock as he gives up his life to save the ship, if we knew that there was a backup copy of him in the database. Another problem with the backup scenario is that the writers would then also have to deal with the social ramifications of it being pretty easy to create multiple copies of a single person. Hopefully there'll be some future ordinance against that sort of thing.

The introduction of non-human sentience presents another inconvenience. It makes it harder to write interesting stories when the humans are always outnumbered by intelligent androids, all having the same rights and privileges in the pursuit of happiness. Fortunately for the script writers, although such androids exist in the twenty fourth century, apparently they remain very difficult to manufacture, which cleverly circumvents the problem of having to deal with lots of them.

What will the world actually be like in a few hundred years? On the positive side, *Star Trek* portrays a hopeful vision of the future; a world without hunger or want.

"Material needs no longer exist," says Captain Picard.[5]

Yet, human frailties remain prevalent, and the occasional corrupt admiral still shows up in the Starfleet ranks from time to time, wreaking all kinds of havoc. I suppose if things were to get too civilized in the future, it could adversely affect the show's ratings here in the present. In "The Measure of a Man," we learn that our civilized future still has difficulty figuring out if intelligent machines are truly sentient, and entitled to the same civil liberties as the flesh and blood members of society, or whether they should simply be regarded as a form of property.[6] The dilemma worked reasonably well when staged in 1989, but how likely is it that we'll be grappling with these kinds of questions three hundred years from now?

There are many examples in science fiction, of robots suddenly and unexpectedly becoming sentient. What is the likelihood of this actually happening? I would think that the probability of self-awareness spontaneously occurring in my desktop computer must be pretty close to zero. Admittedly, some of its many applications do seem to emulate the behavior of the mind, at least a little. It can play chess, for example, which was once something that only a human brain could do. However, in reality, the machine is just running through a series of instructions, albeit with some degree of efficiency. The various decision points in the software are what give the computer the ability to think its way through a problem. If you listen to software developers discussing their work, you will often hear them personifying the source code in this way.

"You see, in this piece of code, the system thinks that the shipment has been sent, so it decides to notify the customer."

The system isn't really doing any thinking, or decision making, in the way that humans do. It simply operates on the available data in the manner exactly prescribed by the software. No other course of action is possible, and no independent thought is necessary.

The human brain is a highly complicated piece of work, developed over millions of years in small incremental steps under the guidance of natural selection. While in many respects we've been able to outpace nature with technology, our computers to date have had fairly limited cognitive potential. They are great at remembering details, much faster at performing calculations, and they can help us to look for patterns in large amounts of data. Like most of our mechanical wonders, computers can patiently work through repetitive tasks without getting bored or restless. Much progress has been made under the broad category of artificial intelligence, including systems that recognize faces and interpret human language. But while science fiction enthusiasts continue to hold out hope for a general purpose electronic, or positronic brain, our technology to date has not been up to that particular challenge.

Generally speaking, human engineered solutions, as useful as they are, cannot hold a candle to the intricate and complicated workings of nature. Consider flight, for example. Nobody has yet been able to create anything as marvelous as a bird. The control surfaces of an airplane look very crude when compared to the dynamic forms of feathered wings, but this crudeness doesn't prevent us from flying at extreme speeds and altitudes – much faster and higher than any of nature's creatures. Alongside the graceful subtleties of bird flight, ours is by comparison more of a brute force approach. Yet, for the purposes of getting airborne, our technology is more than adequate. There are many accomplished aerospace engineers who can do their jobs quite well without delving into the microbiology of birds. Building an artificial brain, on the other hand, may require closer attention to nature's blueprints.

The blueprint for creating a brain can be found in DNA. A full copy of your personal DNA resides inside every one of your cells, along with all of the machinery used to access the instructions, or genes, for building proteins – nature's construction materials. Proteins are built to task, and are used to perform a multitude of functions, including the construction of new biochemical machinery. Like us, cells have the ability to go forth and multiply, and they do this via a process called cell division, in which a cell divides into two new cells. Whenever this happens, a whole new cellular factory is copied into each of the progeny, including a copy of the entire gene library. This is how an organism grows. Each of us starts out as a small glob of undifferentiated cells which then branch out, and differentiate into cells that will perform specialized roles in the body, such as skin cells, blood cells, or muscle cells. Those used to build brains are called neurons, a unique class of cells with specialized signaling capabilities.

As you would expect, different features require different sorts of building materials, or proteins, which means that cells need to be precise about when and where specific genes are activated, or expressed. Genes expressed during embryonic development, for example, can trigger a cascade of additional genes which in turn produce the proteins that govern anatomical growth. Decisions concerning which genes should be active or inactive often depend upon chemical signals received from other neighboring cells. On the basis of such signaling, cells can figure out whether they are supposed to grow into an arm, for example, or a foot, depending upon where they are located with respect to their neighbors.

Presumably then, the installation instructions for building a brain are contained within some subset of the 25,000 or so protein-coding genes that make up the human genome. A typical brain contains maybe a hundred billion cells, or neurons, all intricately wired together into a complex communications network. A single neuron may have as many as fifteen thousand connections to other cells. At first glance, it's difficult to fathom how such a complicated structure can be assembled using a relatively small set of genes. A clever designer would probably plan out the circuitry in advance and then install everything carefully, making sure that all the wires were properly soldered together. Nature, on the other hand, prefers more of a trial and error approach. In the nascent brain, before a baby is born, neurons reach out to one another to start networking. By the time the newborn arrives, there are now many more neural connections than will eventually be needed. As the brain begins to soak up sensory information about its surroundings, functional neural circuits begin to form, and redundant connections are pruned away; a process sometimes referred to as "neural Darwinism." In this way, the physical architecture of the brain grows and adapts in response to its early childhood experiences.

This means that nature doesn't have the luxury of an instruction manual in which everything is spelled out to the last detail. Instead, most of the specifics seem to be determined on the fly. Thankfully, the process isn't quite as haphazard as that might imply. Far from being a free-for-all, there are rules governing the growth of neurons, and the kinds of networks they can form. In fact, as Stephen Wolfram explains, you can get a surprising degree of complexity from the repetitive application of a very minimal set of instructions. He provides more than a few examples in *A New Kind of Science*, illustrating how fairly basic computer programs can easily produce sophisticated patterns.[7] Computer programmer Craig Reynolds shows that the same approach can be used to model aspects of nature that traditionally haven't been easy to represent mathematically. It's hard to imagine a simple way to describe, for example, the constantly changing pattern made by a flock of birds as they swoop around, somehow managing to avoid midair collisions. Reynolds developed a program, (appropriately named *Boids*), that generates a remarkably realistic simulation of this flocking behavior by assuming that the motion of each bird is guided by just a few simple rules. The technique works equally well for schools of fish, and swarms of insects.[8]

In similar fashion, computers are helping to unravel some of nature's most convoluted forms, using the nested patterns of fractal geometry to represent the jagged contours of coastlines, for example, or the branching structure of trees, blood vessels and neurons. The most noticeable thing about fractal images is that, as you zoom in and out to different levels of magnification, you get the same pattern duplicated, over and over again. Perhaps not surprisingly, we see something similar in the shapes produced by nature. The branching pattern at the base of a tree, for example, resembles the pattern at each of the nodes, all the way up to the smallest branches at the top. This approach makes good sense, as it allows nature to be conservative with the genetic code. All kinds of elaborate biological systems can be produced by recursively invoking common genetic subroutines at various branch points.

Computer graphics expert Loren Carpenter was one of the first to show how fractal based software could be used to generate convincing representations of natural landscapes. His work was featured in the movie *Star Trek II*, with a sequence that showed the terraforming effects of the Genesis device. It was the first movie to use a computer generated fractal landscape. Computer animation has since flourished, with continuous improvements in the ability of the software to replicate natural effects. Film makers are no longer restricted to using sets made from cardboard and polystyrene. Almost anything imaginable can be brought to life, from dinosaurs to fairy tales. This beautiful new cinematography is an indication that we are getting better at understanding and copying nature's designs, and so it seems likely that computer simulations will also play a role in unraveling the complexities of the human brain. Perhaps, in due course, the software developed for this purpose might even be used to help build the first android brains.

The *Boids* program provides a good example of a concept called emergence. The idea is that you start with a set of precepts, such as the rules governing the individual motion of a virtual bird, and you end up with something new and unexpected; in this case, flocking behavior. Likewise, if you take hydrogen and oxygen, two invisible gases, and you combine them together, you get something liquid as a result. The liquid character of H_2O is an emergent quality. It is a natural property of water, and yet not necessarily something you would have predicted from the outset. Individual H_2O molecules could hardly be referred to as wet. In the same way, consciousness and self-awareness can be considered emergent properties of the brain. We don't intuitively expect that from a gelatinous mass of interconnected neurons, sentience will somehow emerge. Yet, as science writer James Shreeve puts it, "The mind is what the brain does."[9] To be more precise, consciousness is something that only a certain kind of neurological activity can produce, since obviously there are such things as unconscious brains. So what kind of activity are we talking about?

The neuron is the simplest element of the brain, in the same way that a transistor might be regarded as the most basic component of a computer. The main body of a neuron is in essence a bag of liquid called cytoplasm, in which lots of little protein machines can be found floating about. Among these tiny entities are the mitochondria, miniature factories that produce energy for the cell. Like other cells, the neuron also contains a nucleus, the inner fortress, or command center, within which we find a copy of the gene library. Among the two hundred or so cell types of the body, neurons are perhaps the most personal to us, the cells to which we are most affiliated. For not only do they carry information to and from our brains, neurons are also responsible for what goes on inside the brain itself. They are what constitute the gray matter that my granddad used to talk about.

If you've ever examined the shape of a neuron, you'll probably have noticed that it has a long wire dangling from the cell body. This is called an axon, and it carries the output signal from the neuron. Whenever a neuron wants to send a message to another neuron or to a muscle cell in some remote corner of the body, it transmits an electrical pulse called an action potential, which travels out along the axon to other connecting neurons. In the peripheral nervous system, the axons are coated with insulation, and bundled together like telegraph cables, down through the spinal column, and all the way out to the fingers and toes. The insulation, called myelin, helps the action potentials to travel farther and faster. We refer to these cables as nerves.

Axons come in various lengths, depending upon the function of the neuron, and they may have branches, meaning that the same signal ends up going to more than one recipient. If I were to prick my finger, as sometimes happens when I get caught in the rose bushes, sensors in my finger would relay the pain signal to my brain, though a series of connected neurons. When I decide to move my arm, my brain sends a signal in the opposite direction, through a set of motor neurons that cause the appropriate muscles in my arm to contract. Our bodies can sometimes act on sensory information before it gets all the way to the brain, through what are called reflex actions. This is made possible by cross wiring through the use of relay neurons, resulting in shorter neural circuits and faster responses.

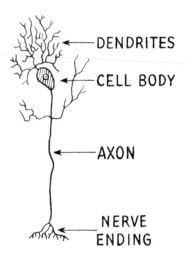

Figure 5. Basic parts of a neuron.[10]

Neurons must be able to talk to other neurons, so there needs to be a way to string them together. In addition to the output wire, or axon, each neuron also has a bunch of input threads, called dendrites, which grow from spiny outcrops on the cell body. Brain cells can have a profusion of such thread-like appendages, spreading out like the thin branches of a tree, thus allowing a single neuron to receive inputs from thousands of other neurons. Caltech scientist Christof Koch describes these dendritic trees as resembling a lot of "fuzzy hair."[11] Connections between neurons are called synapses, and generally speaking, neurons connect where axon meets dendrite, although it is also possible for an axon to make a connection directly with a cell body, or with another axon.

If you were to look very closely at one of these synapses, you would notice that the axon and dendrite don't actually touch, but are separated by a small gap, called the synaptic cleft. So, when one neuron talks to another, how does the signal bridge this gap? The somewhat surprising answer is that when an action potential reaches the end of an axon, it triggers the release of messenger molecules, called neurotransmitters. These signaling agents then set sail across the synaptic cleft, like little boats, to arrive at their docking stations on the other side. Once a neurotransmitter has docked with an appropriate receptor, this action opens a miniature door in the cell membrane, allowing electrically charged particles called ions to enter the cell. This brings about a slight change in the electrical charge inside the receiving neuron. If multiple input signals are received at the same time, the buildup of electrical charge may surpass a certain threshold, causing the postsynaptic neuron to fire a signal of its own, sending an action potential down its axon to the next neuron.

You could be forgiven for saying that the use of neurotransmitters seems like an overly complicated way for neurons to communicate, although maybe it's not that surprising, given the prevalence of this sort of chemical signaling between other kinds of cells in the body. Surely it would speed up the thought process if we could use a system that was entirely electrical, instead of employing chemicals to swim across gaps and open up ion channels; a process that can take a millisecond or less to complete.[12] As it happens, electrical synapses are actually used in some cases, and as you'd expect, they are much faster and also simpler than the chemical versions. Simpler is not always better, however, and for our minds to function properly, it seems we need the more complicated synaptic behavior that you get from having different kinds of neurotransmitters and receptors.

Neurotransmitters fall into two broad categories – excitatory and inhibitory. Excitatory signals will increase the likelihood that the target neuron will fire, whereas inhibitory signals have more of a calming effect. For example, a neurotransmitter known as GABA (gamma-aminobutyric acid) plays an inhibitory role, which means that a GABA deficiency could allow neurons to get over excited, thus contributing to a feeling of anxiety. Drugs such as Valium can counteract this anxiety, because they have the right molecular shape to engage with the GABA receptors.

Koch points out that, as with any electrical system that involves positive feedback, there is a danger of "runaway excitation," which in the case of the brain can cause epileptic seizures.[13] To keep these electrical oscillations under control it is thus important for the brain to establish a dampening system, which means maintaining a proper balance between inhibitory and excitatory signals.

So, while the neuron may in some ways be analogous to a transistor, we can see that there are some significant differences between the two. In a computer chip, a transistor is essentially a little switch, that turns on or off in response to a reference voltage applied to its gate electrode. In this way it creates the ones and zeros of binary logic, used for digital computing. For a neuron, however, the decision to flip the switch is based on a summation of all of the inputs it receives, both excitatory and inhibitory. With that in mind, even the simplest element of the brain looks less like a transistor, and more like a computer in its own right.

Although most diagrams of the brain give the impression that the brain is made up of discrete sections, in reality there are no independent parts to the brain, as everything is intricately connected together. In general terms, though, the uppermost cerebral cortex is responsible for processing sensory input and for executive functions such as decision making, motor control, and planning for the future. If you ever see a movie categorized as cerebral, you'll know you're looking at a subtitled, foreign production, with some incredibly obscure plot that makes your head hurt.

A little deeper in the hierarchy are the so-called limbic areas that deal mostly with instinct and emotion, and below that is the thalamus, often referred to as the gateway to the cortex. Without the thalamus, the cortex isn't able to work properly, and you end up in an unconscious, vegetative state. The lowest section of the brain manages automatic tasks, such as coughing and sneezing, as well as the regulation of important functions such as heart rate and breathing.

The brain is thus organized a little like a typical office building. The managers reside on the top floor, where they get the best view of the city, while building operations and maintenance are housed on the first floor, or in the lower levels, along with the furnace and the water pumps.

Most of us are pretty familiar with the brain's distinctive football shape. I have a vague childhood memory of my mother holding a cauliflower, and pretending it was a brain. More recently I came across a website that showed how to make a Halloween brain dish, using steamed cauliflower, covered with a beet-based vinaigrette sauce to give it a nice blood stained appearance.

In actuality, the cerebral cortex comes in two flat sheets, a little like rolled-out pizza dough, approximately two millimeters thick. The sheets are folded up tightly so that they can fit inside our skulls, and the folds are what give the brain its convoluted appearance. If you were to look closely at a flattened out cortex, you'd notice that it's made up of layers of neurons. There are six commonly recognized layers, labeled from the top down, with layer one closest to the skull. Our minds result from the activity in this layered cortex.

There are some activities that don't seem to require a lot of conscious thought; things that we can do on autopilot, such as going for a walk, or driving home from work in the evening. But even when we are fully focused and alert, the brain is doing a lot of work behind the scenes, outside of our awareness. According to Koch, we are not even aware of the decision making and planning that occurs in our brains; at least not until we use visual imagery or internal speech to reflect on the outputs of this processing. In that sense, our awareness of the world is always lagging behind reality.[14]

As I watch my daughter play soccer, I would intuitively expect that my eyes are taking in all the action, like video cameras, and conveying images to my brain, where the scene is then projected onto some internal view screen. In reality, the situation is a little more complicated. The light that enters your eye is first picked up by light sensitive cells that are connected to different sets of retinal neurons. Once a neuron sees what it's looking for, it gets more excited and increases its firing rate to more of a rapid fire pattern. These response patterns are sent back along the optic nerve to a relay station in the thalamus called the lateral geniculate nucleus (LGN). From there, the signals are shunted off to an area at the back of the brain called the visual cortex.

Surprisingly, signals are also being sent in the reverse direction, from the visual cortex back to the LGN. In fact, the number of neurons heading back into the LGN is far greater than the number going out. As Koch observes, "Think of a video camera hooked up to a computer with a much thicker cable snaking back from the computer to the camera."[15] This enormous amount of feedback suggests that the brain is doing a lot of signal discrimination and modulation of the sensory data.

It's tempting to refer to the visual cortex as the brain's video card, even though the analogy isn't quite accurate. If you can visualize the flattened sheet of cortex again, what we're saying is that there's a section of this cortex that's responsible for processing all of the data that's streaming in from the eyes. The visual cortex is actually composed of a number of subsections, each responsible for some particular interpretation of the sensory input. In the primary visual cortex (V1), for example, most of the neurons seem to care about straight lines, such as the edge of an object. In fact, these neurons even show selective preference for lines that are oriented at particular angles. Some will get excited about a 45° tilt, for example, while others react more strongly to vertical lines.

If you were to insert a small wire electrode down into the primary visual cortex, it would pass through a set of vertically aligned neurons that all respond to a particular point in the view field, and to a specific characteristic of the image. For example, they might all handle the perception of a straight edge at a given orientation. What you've got then is a tiny column of neurons that deals with a particular visual percept. An adjacent column might, for example, deal with straight lines that have a slightly different orientation. In this way, the retina is mapped out onto the visual cortex, with disproportionately more cortex given over to the fovea – an especially sensitive retinal location.

It's clear that the brain has quite a bit of work to do before our minds can derive any meaning from what the eyes are taking in. Once the preliminary mapping is done, V1 then sends its output to other parts of the cortex for additional processing. For this communication to take place, different areas of the cortex are connected so that they can talk to one another, and there is a certain order to the way in which they are wired together. Layer four is typically used to receive input from other cortical regions, whereas outbound signals, going to motor neurons for example, would use layers five and six.

There is a general hierarchy to the way information is represented in different sections of the brain (not to be confused with the six layers of the cortex). In the visual pathway, the hierarchy is organized more or less from the back of the brain to the front. The basic sensory input comes in via the LGN to the V1 area back of the brain, and these percepts then get summarized and categorized as we move forward to the frontal areas.

For example, sensory data from V1 is fed into another part of the visual cortex, called the middle temporal area (MT), which contains neurons that obsess over signs of movement in specific directions, such as an object moving across the field of view from right to left.

In the frontal cortex, information is represented at the level of concepts that are more general, or symbolic in nature, and are not tied to any of the specific sensory feeds. The communication doesn't just flow in one direction, as there are also feedback cables from the frontal regions, reaching back into the lower levels of the hierarchy. It is as if we're getting signals from upper management saying "This is what I expect," while the plebs in the basement are saying "Here's what's actually happening."

Even though different regions within the cortex have specialized duties to perform, their neuronal wiring looks pretty much the same, meaning that the same fundamental architecture is used throughout the cortex. This helps to explain the plasticity, or adaptability of the brain. Experiments on animals have shown that parts of the brain can be retrained to take on new responsibilities. Neuropsychologist Alvaro Pascual-Leone, of the Beth Israel Deaconess Medical Center in Boston, has shown that this retraining also takes place in humans, and that it can happen over a period of just a few days.[16] This is important for patients who've experienced brain damage, since it means they can potentially recover some of the functionality that was lost. We saw this happen in the case of Congresswoman Gabrielle Giffords who was shot in the head on January 8, 2011. In the months that followed, the congresswoman made a remarkable recovery and gradually regained the ability to speak and to walk.

The idea is that neighboring parts of the brain can compensate for the damage, learning to perform the duties of the missing or injured neurons. Theoretically, you could rewire the auditory inputs, connecting them to the visual cortex, for example, so that the neural circuits that were previously used to process visual signals would instead be commandeered for hearing.

In Pascual-Leone's experiment, a volunteer was blindfolded for several days, so that her visual cortex was deprived of any sensory input. She was then asked to use her fingers to read and compare a series of braille patterns. After a while, her brain began using its now quiescent visual cortex to process the new information coming in via the sense of touch. Since the visual cortex had nothing better to do, it was free to take on some additional work.

The concept might not work as well across all cortical areas. For example, you could run into some problems if you connected visual input to the motor cortical areas that have a relative sparsity of neurons in layer four. That's because the motor cortex is usually concerned with sending outbound signals to muscle cells, so there was never any need to develop a rich input layer, unlike the visual cortex which receives a massive amount of sensory input from the eyes.

I should mention that there are a number of separate pathways that sensory data can take. One such pathway bypasses the cortex and leads directly to the amygdala, the emotion center of the brain. Since there's less processing involved, this is useful when faster response times are important. If I'm driving on the freeway and the car in front abruptly comes to a stop, I need to be able to quickly slam on the brakes without taking the time to think about it. That's where the amygdala pathway comes in handy.[17]

The pathway leading to conscious thought, on the other hand, involves a bit more work at the cerebral level, as sensory data undergoes further processing and interpretation. Koch suggests that the general behavior of the cerebral cortex can be described in terms of neuronal networks in the frontal lobe looking back at the rest of the cortex which supplies all of the sensory input. These frontal networks then assess the data, make decisions, and send appropriate instructions to the motor neurons. We remain unaware of these processes, until the final determinations are served up to become part of our conscious deliberations.

So, as I'm watching my daughter's game, the end result of all this cortical processing is that I can understand what's happening on the soccer field – that is, unless there's an offside ruling. This conscious awareness of what's going on is not something that can be pinned down to a specific place in the brain. Instead, Koch believes that consciousness may involve a coalition of interconnected neurons, all chattering together. At any given time there may be several such neuronal coalitions representing different percepts, happenings or imaginings that are competing for attention. As the activity of particular coalition reaches a certain threshold, it takes center stage, while another fades into the background.[18]

Koch may be in a better position than most to speculate on the nature of consciousness, but even he admits that it is impossible to definitively say what it is. Interestingly though, we have amassed enough clinical data to say what it is not. Along with fellow neuroscientist Giulio Tononi, Koch points out that consciousness does not require attention, memory, self-reflection, sensory input, or the ability to move. Neither does it require emotion. Damage to the frontal lobe can leave some individuals in an emotionless state, but however dispassionate they may be about their situation, they are still perfectly capable of being conscious.[19]

How do researchers know what happens inside the brain? The most effective approach is to insert a small wire electrode into the cortex, so that we can listen to the neural activity at that precise point. We can also use the electrode to apply a small electric jolt to stimulate the neurons, and see what it does to the individual. Most of the time, it's not appropriate to start poking around inside a human brain, but sometimes there are medical reasons for doing so. On such occasions, researchers can learn a lot about the functioning of the brain by communicating directly with the patients who are having their heads examined. It's even possible to keep a patient awake to assist the surgical team during a brain operation. The surgeon's probe is guided by feedback from the patient, so that a tumor can be removed while minimizing damage to the surrounding neuronal circuitry.

Patients suffering from seizures often have their brains monitored via implanted electrodes, so that doctors can zero in on the problem areas. In such cases, scientists are sometimes able to use these embedded electrodes to learn more about the functioning of the brain. At the UCLA David Geffen School of Medicine, under the supervision of neurosurgeon Itzhak Fried, a team from Koch's lab recorded the neural firings that were produced as patients looked at images of various celebrities. In the case of one patient, they noticed that one of the electrodes recorded a signal whenever the patient saw an image of the actor Josh Brolin. Koch describes how a different neuron fired in response to the "iconic scene of Marilyn Monroe standing on a subway grill, trying to keep the wind from blowing her skirt up."[20]

In an interesting demonstration, the researchers used a computer to determine which celebrity the patient was thinking about, by detecting activity in the corresponding neuronal circuit. The patient was presented with the two images superimposed on a display screen. When she focused on Brolin, the computer picked up the corresponding signal from her brain, and made that actor's image appear more prominently on the screen, while Monroe's image started to fade. In this way, the patient could control what appeared on the screen just by using her thoughts.

Although the computer can figure out which image the patient is looking at, it cannot know exactly how the celebrities are represented in the brain. Without help from the patient, there would be no way for us to make any sort of correlation between neurons and memories. There are no clumps of neurons bearing any special resemblance to Josh Brolin or Marilyn Monroe. By analogy, we are accustomed to the fact that a video recording can be stored in either digital or analog form as a pattern on magnetic tape, or as a sequence of dots on an optical disc. You can look at a Harry Potter DVD through a microscope, but you won't see any images of Hogwarts. Our computers and video cameras store information in a form that's meaningful to them, and we can then use the same machinery to playback the images in a way that's meaningful to us. The same is true with brains. Even if we had the technology to watch all the neurons firing, and examine all the synaptic patterns being invoked, we'd still have difficulty translating these patterns into something comprehensible.

As with computers, there is a distinction in brains between short term and long term memory. Short term or working memory is the equivalent of what's called Random Access Memory (RAM) in computing. If you unplug a computer before it has a chance to store what's in its RAM, the information is lost. Likewise, if you get hit over the head before you commit something to long term memory, it also gets lost.[21]

The brain can hold an active concept in working memory as a pattern of firing neurons, and it keeps the information in its buffer by repeatedly firing the same neuronal collection. Long term memory is the equivalent of writing the information to a data storage device. To achieve this, the brain must have a way to preserve the neuronal patterns, so that they can later be recalled when needed.

You can think of a neuronal pattern as a set of interconnected neurons arranged in such a way that if one neuron fires, the others will also fire.

In the words of neuroscientist Carla Shatz, "Neurons that fire together wire together."[22] Our memories are stored using a network of such patterns, distributed across different areas of the cortex for extra resilience. Just as business owners store their accounting data in separate physical locations across the country; this makes it less likely that any given memory will be completely lost.

How then do neuronal networks manage to preserve memories in long term storage? For this to happen, the patterns representing memories must somehow be reinforced. One plausible way to accomplish this would be to increase the responsiveness of the synapses, or connections, along the neural pathways that form the pattern. At the synapse level, this entails the recruitment of special receptors[i] to bring about a chemical strengthening of the synaptic connection. Whereas a computer uses a laser to burn its memories onto a disc, the brain does the same thing by sensitizing and desensitizing neuronal synapses. Synaptic sensitization is called long term potentiation (LTP), whereas desensitizing is known as long term depression (LTD); not to be confused with … well, long term depression.[23] Studies conducted by researchers at the University of Illinois showed that memories can also involve the formation of new synaptic connections. Experiments showed an increase in dendritic branching in the brains of rats that learned how to perform new skills in laboratory tests.[24]

While memories can be stored in different places, there is one special area of the brain that is responsible for recording those memories, and it's known as the hippocampus. If you ask a scientist to describe what the hippocampus does, you'll probably hear the story of Henry Molaison, a man who had that part of his brain removed in 1953, and who was known for the rest of his life as H.M. to protect his identity. The procedure was carried out by neurosurgeon William Scoville, in the hope that it would cure H.M.'s epilepsy. It did, but it also rendered him unable to form any new memories. He could remember his life before the operation, and he could learn new skills, such as mirror writing, but he had no idea how he had acquired these skills. You could have a conversation with him, but if you were to leave the room and return ten minutes later, he wouldn't recognize you.

---

[i] NMDA receptors, in particular, are implicated in memory encoding via LTP.

Even as his body grew older, in H.M.'s mind it was always 1953, and he was the same twenty seven year old man until his death, in December, 2008. His unfortunate situation demonstrates that while memories do not reside in the hippocampus itself, this critical section of the brain acts like a sort of central switchboard, through which memories are created. The fact that H.M. was able to recall things that happened earlier in his life implies that other parts of his brain were involved in retrieving previously saved memories. His knowledge about the pre-1953 world seemed pretty intact, particularly with respect to information from the public domain, and yet he was unable to remember specific events from his personal history. He couldn't tell you about something that happened on his tenth birthday, for example.

This would imply that the amygdala-hippocampal region must also play also play a role in memory retrieval – specifically in the retrieval of episodic memories; things that happened on particular dates, or in the context of some event. This may be why our sense of smell appears to be particularly effective at evoking emotionally potent memories, since it is the only sense that is wired directly to this part of the brain.[25] H.M.'s ability to learn new motor skills provided confirmation that procedural memories are recorded via a different route – one that doesn't involve the hippocampus.

In 1985, British musician Clive Wearing suffered a viral infection that caused extensive damage to parts of his brain, including the hippocampus. The resulting memory impairment is so severe that he has effectively lost his identity. Sometimes he may not even have enough working memory to finish a response before forgetting the question he was trying to answer. The only person Clive recognizes is his wife, whom he adores. He is always delighted to see her, even when she has only been out of sight for a few moments.

Such examples underscore the close relationship between individual identity and the function of memory. It is the collection of synaptic patterns inside your head that defines who you are and that distinguishes you from someone else. Deprived of this repository of memories, you cease to exist. Until recently, it was difficult to see how we might someday be able to repair the sort of damage suffered by H.M. and Wearing, yet that seems to be what scientist Theodore Berger has in mind. Berger's team, at the University of Southern California, is working to develop an electronic chip that could be used to bypass a damaged hippocampus. This artificial hippocampus would be fixed to the outside of the skull, but would have interface wires extending into the brain, to connect up with the neurons on either side of the damaged area.

Other kinds of brain prostheses are already in use, such as cochlear implants that help deliver auditory signals, but these are used to augment sensory function rather than cognitive abilities. We often hear that our minds are influenced by technology, but with one of these chips installed, that would quite literally be the case. Although we are unlikely to see artificial brains on store shelves any time soon, it's remarkable that we are already working on a replacement for part of the brain that plays such a critical role in shaping one's identity. We are finally getting to know ourselves, by using our brains to learn how brains work.

"Consciousness is not only important," argues UC Berkeley philosopher John Searle, "it is the condition of the possibility of anything at all being important. Remember, something is important to you only in so far as you are conscious. . . . I think consciousness is not only important, it is so to speak the most important thing because all other important things depend on consciousness for their importance."[26]

It sounds quite profound, but as we uncover the science behind what makes us human, this knowledge will surely have a bearing on what we will become. There is already much speculation on what the future may hold, much of it in the realm of science fiction. Some stories predict that we will upload our minds into machines, and use this as a way to escape death. Others suggest that we will build intelligent androids, such as *Star Trek*'s Lieutenant Commander Data. These are concepts that have a lot of entertainment value, but they also present some interesting perspectives that will be important to consider in the continuing quest to understand our fate.

5

⌘

Synthetic Dreams

My favorite science fiction characters were robots. The first to really catch my attention was Robby the Robot, from the movie *Forbidden Planet*, but I also became quite attached to the one with the inspirational name "Robot," that featured in the *Lost in Space* TV series. I should point out that, apart from my interest in the robot, I was never a big fan of the show. The most impressive robot I've seen in real life, so to speak, was built by Honda and goes by the name Asimo.

During a recent family trip to Disneyland, we were treated to an entertaining demonstration in which Asimo climbs a few steps, plays around with a soccer ball, and even manages to run across the stage. A slow motion camera confirmed for us that both feet actually left the ground as it sprinted along. I guess this was the first time I'd seen a robot that behaved like the ones I'd grown accustomed to from the movies, although there was something missing. As well as having extraordinary balancing skills, I suppose what I've been holding out for is something that has a lot more human-like intelligence – maybe something like the HAL 9000 from *2001: A Space Odyssey*, but which also has the ability to run up and down stairs. Is that too much to ask for?

I'm now a middle-aged man and we still don't have a HAL 9000. Interestingly, my kids seem to have inherited my fascination with robots. My daughter now has a small version of the R2D2 robot from *Star Wars*, and it trundles around the house with its own little personality. I'm always amused whenever my wife inadvertently starts talking to it. She's in the middle of an altercation with the kids when R2 starts acting up. "What's the matter with you?" she asks the droid.

About twenty years ago, an engineer assured me that within a couple of decades, computers would have the same number of connections as the brain, so at that point we'd start seeing androids and the like. Apparently the only difference between computers and brains was the number of connections, whatever that meant. Now that I've gone to the trouble of learning a little more about brains, I'm not so sure he was right. It seems a bit of a stretch to start promising to build computers with a brain-like ability to produce sentience, when we still don't know how the brain does it.

The unfortunate reality is that you cannot just string a bunch of neurons together, like holiday lights, and expect that consciousness will materialize. Nor should we expect that digital computers will suddenly become self-aware, once we reach a certain critical mass of transistors. As we've seen, our minds exist only because of the very specific ways in which our neurons are connected, and because of the kind of activity that this architecture supports. We do not yet have a detailed working model that adequately describes the neural basis of the conscious mind.

The task is challenging, since the human brain is a complicated machine, and one that is not easy to examine while it is functioning. Nevertheless, for a computer to be sentient it's reasonable to assume that it would have to be able to do what the brain does, and we know enough already to be able to draw a sharp distinction between the biological system that houses the mind, and the information technology with which we are more familiar. As much as we like to draw analogies between computers and brains, they operate via fundamentally different principles.

Our computing machines do very little when left to fend for themselves. Without software systems, it is difficult for them to perform any useful function. Although software comes in a variety of flavors, it all ends up being reduced to a series of simple binary code instructions which can be interpreted by a computer processor. Processors are the brains of a computer. They reside on small, flat silicon chips, usually no larger than a few square centimeters, and are thus considerably smaller than a human brain. Each processor has a series of data storage locations called registers. You can think of a register as a row of simple transistor circuits called flip-flops. A flip-flop is what's known as a bistable device, which means that it can be set in one of two positions, allowing it to hold on to a single bit of information.

When the processor is ready to do some work, it fetches the first instruction, and dumps it into one of its registers. What this really means is that the flip-flops in the register get flipped on, or flopped off to match the binary pattern in the instruction. So you end up with a sequence of flip-flops in the on and off positions, which programmers like to think of as representing ones and zeros. Using the binary number system, this means that a sixty four bit register can represent any number you can think of, up to 2^{64}, which is a big number.

Each instruction is, in fact, a binary number that is parsed into different segments. One such segment represents the command itself, whereas other segments may point to the addresses of data storage locations in memory. The data from these locations can be loaded into additional registers. Depending upon what it was instructed to do, the processor then performs some basic binary calculations, simply by manipulating the flip-flops in its registers. That's essentially all that a processor does. It's really just a glorified abacus.

Processing continues until all the instructions have been executed. The good news is that processors are very fast. For example, the processor that I am using to write this chapter is running at 3.2GHz, which means that it's getting through about 3.2 billion instructions every second.

A computer program is a list of instructions that can be executed by a computer. The processor itself can only deal with instructions that are written in the binary language of machine code; that is, streams of ones and zeros. However, you can easily devise a set of more complicated commands, as long as you write a program that can translate these commands into machine code. The simplest such translator is called an assembler, and, it is written in machine code. Each processor variant will have its own assembler, tailored to the architecture of the machine. Once you have an assembler, you can then write programs using assembly language, which is better than having to deal with lots of ones and zeros. Assembly language is thus one step above machine code. And once you've gotten this far, there's no reason to stop. You can continue to invent yet more sophisticated programming languages, as long as you build translators, or compilers, that can convert your new language into assembly, or machine code. The resulting collection of programs is called software, as distinct from the physical computer machinery, or hardware.

You can see how this software can grow and evolve over time, as additional layers of complexity are built upon basic programming platforms. To begin with, we needed software to control the essential operating functions of the machine; writing information to disc, managing programs, files and memory, or displaying computational results on a screen. This led to the development of operating systems, such as UNIX and Windows. Networking software was then introduced to allow computers to talk to one another.

A few decades ago, most programmers could tell you exactly how their software was expected to behave. That's because in those days, if you wanted your program to work, you had to tell the computer everything you wanted it to do. Today's developers, on the other hand, rarely write anything from scratch. Software is developed in a more modular fashion, allowing different modules, or objects, to communicate with each other via messaging systems. To introduce new functionality, developers can extend existing software to perform additional tasks, or make use of services available over the Web. In many cases, the code is developed by computers themselves, since we now use software to create software. The result is a world in which software is ubiquitous; a world in which automated intelligence facilitates the machinery of human culture.

Our culture is driven by an enormous quantity of information that lives outside our heads. With the advent of computers, we began to digitize this human world, starting with the adoption of the ASCII and EBCDIC codes to represent the letters of the alphabet with ones and zeros. Other coding systems were later introduced to handle music, pictures and video. Film and television were converted to new formats and huge libraries of human knowledge and artistic expression have already been digitized.

Converting everything to numbers helps to preserve the content of civilization, as well as making it more accessible. Through the expansion of the internet, a vast planetary database is now available to the masses. This global networking has increased the potential for human collaboration, and provided new avenues for creativity and self-expression. It has changed the way we do business with one another, and how we interact socially, to such an extent that a significant portion of our cultural activity is now conducted in the language of computers. The incredible fact is that all of this information technology, and the richness that it brings to our lives, is somehow made possible by processors fiddling with flip flops.

It is a credit to human ingenuity that, within just a handful of decades, so many wondrous applications have emerged from this relatively simple architecture. It is not, however, the sort of technology that can produce conscience minds, explains UC Berkeley philosopher John Searle. While computers may sometimes appear to be conscious, in reality they are just plodding through sequences of instructions, and blindly manipulating symbols which they do not understand. This is not how brains work, says Searle, which is why our computers can never be sentient. He admits, however, that someday it may be possible to build a sentient machine. After all, the existence of the human brain is proof that such machines can exist. However, any such sentient machine would need to have, in his words, the same "causal powers" as the brain to produce consciousness.

In a sense, our generation is maladapted to the task of figuring out what consciousness is. We have been trained to digitize everything, and to think in terms of software. Brains, on the other hand, do not contain software. There is no separation of memory from processing, there are no flip flops or registers, nor are there layers of explicit instructions that get executed sequentially by the brain. Instead the brain is a system that evolves as it learns, is implicitly driven by external and internal events, and uses an extraordinary degree of parallelism in the storage, processing and retrieval of information.

Nonetheless, there are still plenty of very clever individuals who persist in referring to the mind as software, running on the hardware of the brain. Furthermore, as evidenced by the discussion topics at IBM's Almaden Cognitive Computing Conference in 2006, there are still many scientists who argue that it should be possible to build a conscious machine using a traditional computing approach – a software realization of consciousness.

It is an argument that seems to make sense from a practical

standpoint. Some advances in computing do seem to have brought us closer to synthesizing consciousness. For example, there is a branch of computing that deals with what are called neural networks. These networks do not involve real neurons. Instead, the programmers have devised a way of performing certain kinds of computations using software simulations of neurons. They use weighting algorithms to represent the relative strengths of the connection points between the virtual neurons, just like the long term potentiation of real synapses, except much simpler. The software versions don't try to model the effects of the various different neurotransmitters, for example.

Nevertheless, the technique has been successfully used to help build software that can recognize faces, and interpret speech. Other systems have been used to build robots with advanced motor coordination and mobility. Asimo, for example, is a very likeable robot that can walk up and down stairs with ease, and can run at nine kilometers per hour. In 2011, NASA sent a robot called Robonaut to the International Space Station, as the first step in an effort to relieve astronauts from mundane operational tasks.

If we're already building systems that behave as if they are conscious, then isn't it only a matter of time before we actually replicate consciousness in one of these machines? Inventor and futurist Ray Kurzweil says the answer is yes, and that it will happen very soon. He believes that once these machines are ready, we'll be able to start uploading our minds into them, and that "the end of the 2030s is a conservative projection for successful uploading."[1]

> We will be software, not hardware. . . . As software, our mortality will no longer be dependent on the survival of the computing circuitry.[2]

Searle, however, points out that there is a difference between simulation and duplication. We can build machines that can do some of the same things that humans do, and perhaps even do a better job. They can do these things, however, without being conscious. You might ask if it matters whether a machine is conscious, and in fact most of the time, it probably doesn't, as long as it does the job it was supposed to do. A weaponized robot that appears to have the ability to make autonomous decisions may be just as dangerous, regardless of whether or not it is truly conscious. As a side note, despite the enormous sums that the Defense Advanced Research Projects Agency (DARPA) spends on cognitive science research, it's difficult to see why the military would want a fully autonomous, sentient android, given that they spend so much time and effort trying to get rid of autonomy in humans. The one thing that my dad most disliked about his brief military service was being constantly told what to do, and having to say "Yes Sir," all the time. If they have difficulty with people like him, why would they want to introduce the possibility of a robot that doesn't do what it's told?

In the 1950s, British mathematician Alan Turing devised what became known as the Turing Test as way to approach the question of whether computers have the ability to think. The idea is that you get an independent observer to engage in a textual conversation with an unseen participant, without knowing whether the said participant is a real person, or a computer pretending to be a person. If the observer can't tell the difference, then we might conclude that the computer is actually thinking like a human. I remember one of my college assignments in the 1980s was to use a language called LISP to write some software that could hold a conversation with a human. The idea was to look for certain keywords in the user input, and then have the computer respond with a comment or question related to those keywords.

The same basic approach is used by commercial websites to simulate customer service agents who can respond to typical product support questions. These software entities are known as chatterbots, and while I haven't yet come across a system that can convincingly emulate a human, they are nonetheless a lot of fun to interact with. These kinds of systems are continuing to improve in their ability to do quite well in a standard Turing Test. However, this is by no means evidence of a human-like thought process. Computers can get very good at imitating people, without doing any real thinking at all. This is nicely illustrated by neuroscientists Christof Koch and Giulio Tononi who show how easy it is to stump a computer. Simply ask the question "What is wrong with this picture?" Even a five year old can tell you that it doesn't make sense to replace a keyboard with a flower pot, for example, or for someone to place a chessboard in the oven. But it's very difficult for a computer to draw the same conclusions, thus highlighting an important difference between human and computer intelligence.[3]

The reason has a lot to do with how information is stored. As we learn about the world, our brains focus on the relatedness of things. In the storage and retrieval of memories, context is everything. This allows us to quickly ascertain the importance of any particular fact, because each piece of information is stored in a way that reflects its relevance to everything else that we know. As author Diane Ackerman puts it, "The brain is a pattern-mad supposing machine."[4]

In traditional database systems, however, data is generally recorded as a collection of unrelated bit patterns. Metadata can be used to impose some structure on these patterns, and to build some connectivity between disparate data points, but this is usually done after the data has been collected. Even a rudimentary level of interconnectedness can be very powerful when applied to large volumes of data. We can build systems to help us to recognize a potential terrorist, for example, by matching on a relatively small number of data points. However, these systems have no intuition. It is impossible for them to reason about the world, as we do.

When we ask if a machine is really conscious, the question comes down to whether it feels like anything to be that machine. Ultimately that may be an important question to be able to answer if we ever decide to replicate human consciousness in a machine. Ethically speaking, a conscious, sentient android will be entitled to some respect. But is there any way to answer that question, given its subjective nature? Isn't it the case that the only way we can know if the machine is conscious is to be the machine? Perhaps there is another way. Searle knows that his dog is conscious, not because of the way it behaves, but because its doggy brain works on the same principles as Searle's human brain. Therefore it should have the same "causal powers" to produce consciousness. It may not be the same as human consciousness, but we can at least say that it must feel like something to be a dog.

If we can build an artificial brain that works on the same principles, such that it produces consciousness in the same way that the human brain does, then we can be pretty confident that we're looking at a conscious machine. This is another way of saying that the closer we get to being able to describe the neuronal correlates of consciousness, the easier it will be to recognize sentience when we see it. In that case, engineers of the future might find it relatively easy to determine whether a particular device is as sentient as Searle's dog.

If Searle is correct, then we may conclude that our computers are not exactly on the path to sentience, and if we ever build a thinking machine, we'll need to come up with a very different underlying architecture. In what may be the first steps in this direction, IBM announced in 2011 that they had started testing a new kind of processor, one that more closely models the neural circuitry of the brain. They're called cognitive computing chips, and are smaller and use less energy than regular processors. Big Blue says they'll soon begin showing up in adaptive control systems that can learn through experience and interact more efficiently with the environment.[5]

Even if computers can't become conscious themselves, they can be used to simulate the brain activity that produces consciousness. The most ambitious such attempt is the Blue Brain Project, of which IBM is also a major sponsor. Led by Henry Markram at the EPFL in Switzerland, the project aims to build a complete model of the cortex of a rat. Markram's team started out by completing a detailed simulation of a cortical column, in which each constituent neuron is captured in software, right down to the ion channels in its cell membrane. These columns or "brain chips," as Markram calls them, are then wired together, eventually culminating in the creation of a virtual rat brain. Theoretically, the work could be extended to model a human brain, although a much more powerful supercomputer would be needed.[6]

This kind of modeling is impossible without first being able to get our hands on more detailed information from the brain itself. Perhaps the most popular way to have your head examined is via functional magnetic resonance imaging (fMRI). This type of scanner uses a magnetic field in lieu of harmful radiation, so it provides a safe and non-intrusive way to peer into a living brain. For example, in a series of memory tests conducted by psychologist Daniel Schacter at Massachusetts General Hospital, volunteers were asked to memorize a list of words while having their brains scanned. Using the fMRI images, Schacter was able to determine the brain pattern that corresponded to whether or not something had been committed to memory, which allowed him to accurately predict the words that each individual would remember. Not surprisingly, Schacter notes that each successful memory would light up an area of the brain known as the left parahippocampal gyrus– "one of the regions that HM's surgeon had removed."[7]

In his role as a TV science presenter, actor Alan Alda spends a lot of time in MRI machines, so one assumes it must be pretty safe. Usually we see him lying horizontal, with just his legs visible, and the upper part of his body embedded in the machine. With his head in the scanner, Alda responds to a series of questions, or focuses his attention on a display screen, while researchers observe how his brain reacts. They can see which parts of the cortex light up as he performs these mental exercises. Each fMRI image shows a slice through the brain, and a computer can combine all the slices, from front to back, into a detailed three dimensional model.

While the images look pretty impressive, they are not detailed enough for us to see what the actual neurons are doing. Instead, we can see small localized increases in blood flow which correspond to neural activity in the region. The spatial resolution of the scans is limited to about a millimeter, and it takes a couple of seconds for the blood flow changes to show up, which is not fast enough to catch the synaptic activity that's happening in milliseconds. An MRI scan can prove that the brain is up to something, and we can see where the action is occurring, but we cannot tell exactly what's going on.

Another safe and non-intrusive way to examine the brain is to use what's called an electroencephalogram (EEG), to tune in to the electrical activity of the brain. In this case, we're actually picking up the electrical patterns that are generated as a result of lots of neurons chattering amongst themselves. As with the fMRI scans, an EEG doesn't provide any detail on individual neural pathways, so we still end up with a fairly crude representation of the working brain. Koch likens it to hovering over a ball park in a blimp. You may be able to tell when something exciting happens in the game, from the roars of the crowd below, but it's impossible to make out individual conversations.

To get a closer look at specific neural circuits, we can insert wire electrodes directly into the cortical tissue. In Koch's blimp analogy, this would be like dropping a microphone down into the spectator stands. This is obviously a slightly more intrusive test, and one that I don't remember Alan Alda being subjected to, but with this method, it gets easier to eavesdrop on specific neural conversations, with less noise and interference from the rest of the brain.

Researchers can also use electrical probes to stimulate a particular area of interest, and by talking directly to their subjects they can determine which cognitive abilities are affected. It is easier to do this while working with a human subject, as opposed to, say, a monkey. For one thing, the cognitive abilities of monkeys don't exactly line up with ours, and it's hard for them to understand the purpose of the experiment. Also, when stimulating a monkey brain, you can't expect the same sort of feedback that you'd get from the likes of Alan Alda, for example. On the other hand, Alda is less likely to offer himself up for experimentation.

Usually this kind of testing is only possible with patients that are undergoing surgical procedures, and when it is medically justified. This means that for the most part, researchers must rely on experiments using laboratory animals. Monkeys are generally too expensive to work with, so Koch does most of his research with mice.

In 2003, Brazilian scientist Miguel Nicolelis proved to the world that a monkey brain could be hooked up to an artificial arm. An implanted electrode array was used to monitor the monkey's cortex as it maneuvered a joystick to grasp a simulated object in a video game. These brain signals were then interpreted by a computer which directed a robotic arm to respond appropriately to the monkey's intentions to grab hold of the object on the screen. The artificial arm was thus mimicking the monkey's actions in the video game. The monkey soon realized that it could stop using the joystick, and just control the game directly with its thoughts.[8]

Five years later Nicolelis came up with an even more impressive demonstration. This time a monkey in North Carolina used its brain to control a set of robot legs in Japan. The monkey's brain was rigged up so that its motor commands were sent via satellite to the remotely controlled robot, which the monkey could see on a video screen. As the monkey walked on a treadmill, it was also directing the movements of the robot. Interestingly, when they stopped the treadmill, the monkey continued to control the robot legs with its mind for several minutes.[9] [10] Apart from creating a monkey version of Steve Austin, the ultimate goal, says Nicolelis, is to help people who are suffering from paralysis.

I grew up watching in Brazil the challenge to conquer space, to go to the Moon. I think we can have equivalent challenges right now, a challenge on paralysis, to fight paralysis and hope that in a decade we could basically remove this scourge from our lives. These things can happen now.[11]

In another, somewhat related experiment, a research team led by Ferdinando Mussa-Ivaldi of Northwestern University in Illinois created a different kind of cyborg by using the brain of a sea lamprey to control a small robot, which moved around in response to light. This kind of research can help us how to understand how to electronically interface with a brain, and can also shed some light on how brains adapt to sensory input, and learn about their environments.

The *Harry Potter* approach to brain science is to zap someone's head with a special magnetic wand. Strong magnetic fields can interfere with electrical signals in the neurons, so researchers can use these wands to selectively disable a particular part of the brain to see what effect this has on the subject's mind. The technique is called transcranial magnetic stimulation (TMS).

In an interesting demonstration of this method, scientist Liane Young showed that by disrupting an area called the right temporoparietal junction (RTPJ), you could influence a person's moral judgment. That particular spot in the brain is believed to play a role in helping us to judge the mental state of another person. If we decide, for example, that someone's intentions are evil, then chances are it's the RTPJ that's helping us to make that assessment.

What would happen, wondered Young, if the RTPJ was disabled? So she and her team put together a study in which subjects were presented with a number of scenarios, and were then asked to decide whether the actors in these scenarios were behaving in a morally responsible way. The following is an example:

> Grace and her friend are taking a tour of a chemical plant. When Grace goes over to the coffee machine to pour some coffee, Grace's friend asks for some sugar in hers. The white powder by the coffee *is just regular sugar*. Because the substance is in a container marked "*toxic*", Grace thinks that it

is *toxic*. Grace puts the substance in her friend's coffee. Her friend drinks the coffee and *is fine*.[12]

While everyone seemed to agree that Grace's actions were reprehensible, the people who'd been zapped in the RTPJ were significantly less likely to judge her harshly. After all, nobody got hurt, right? However, if the story was changed so that Grace's friend actually dies, then the difference of opinion was not so pronounced. Everyone then acknowledged that this was pretty bad, regardless of whether TMS had been used on their brains. In other words, if your RTPJ is turned off, you're more likely to base your judgment on the final outcome, and whether any harm was actually done, rather than the fact that Grace intended to do harm.

Yet another way to learn about brains, and probably the most informative from a human perspective, is by studying those unfortunate individuals who have suffered some form of brain damage, either through injury or surgery. Usually, when this happens, the mind is altered in very specific ways, as we saw with H.M. Cognitive functions that are missing, or impaired, can be directly associated with the damaged areas of the brain.

Adding to the mystery of the brain is the fact that it comes in two halves, which are joined together by a thick bundle of neurons called the corpus callosum. Every indication suggests that both halves are, in fact, independently conscious. This extraordinary architecture would suggest that each of us actually has two separate minds working together to appear as one. As remarkable as it sounds, it may be that when my wife says that she's in two minds about something, it might literally be true. It reminds me of the Christian idea that God comprises no less than three separate persons collaborating with each other to form one single entity. The separate minds that comprise the brain are easier to detect when the corpus callosum is severed, as illustrated in this account from Koch of an interview with a split-brain patient:

> When asked how many seizures she had recently experienced, her right hand held up two fingers. Her left hand then reached over and forced the fingers on her right hand down. After trying several times to tally her seizures, she paused and then

simultaneously displayed three fingers with her right hand and one with her left. When Mark [the neurologist] pointed out this discrepancy, the patient commented that her left hand frequently did things on its own. A fight ensued between the two hands that looked like some sort of slapstick routine. Only when the patient grew so frustrated that she burst into tears was one reminded of the sad nature of her situation.[13]

While our understanding of the brain continues to grow, the information that's been collected so far is rather crude and disjointed. We can say, for example, that the neuromodulator oxytocin relates to trust and trustworthiness, or that dopamine is associated with reward pathways.[14] However, the exact neural underpinnings of behavioral and cognitive attributes are still shrouded in mystery. As we've seen, it is impossible to get a really detailed view of the happenings in the human brain, which obviously frustrates the quest to understand the mind. It may mean that, science fiction aside, we can forget about reverse engineering the human brain any time soon. For now, by focusing instead on the brains of mice, scientists such as Koch can tackle the problem in a more methodical way, by examining specific behavioral traits of laboratory rodents and carefully tracing them back to their neuronal correlates.

While we can confidently expect that the next few decades will bring some fascinating insights into the nature of the mind, the rate of progress may be too slow for those who are eager to see human consciousness replicated in machines. Nobel laureate Gerry Edelman cautions us that "in science, ten years is about as much as you can try to look ahead, otherwise you're a crackpot."[15]

Nevertheless, it is tempting to speculate a little on what the future may bring, especially given the potential impact of this area of study. As Searle says, "Consciousness is not only important; it is the condition of the possibility of anything at all being important."[16]

If the mystery surrounding consciousness is unveiled, and if we also succeed in creating artificial brains, this will certainly be a game changer when it comes to our quest for salvation, and human longevity. Suppose, for example, it became possible in the future for machines to access our memories. If that were possible, then perhaps these memories could be stored somewhere, allowing us to keep backup copies of our minds. Jack Gallant at the University of California, Berkeley, provided the first tentative hints that this sort of mind reading might someday become reality.

Figure 6. Mind-reading using fMRI. The top row shows frames from the movie that the subject is watching while inside the MRI scanner. The bottom row shows the computer's attempt to reconstruct the movie using fMRI data from the subject's brain.[17]

The idea behind Gallant's experiment is to have a human volunteer watch a movie while inside an MRI scanner, and then have a computer try to recreate the movie by reading that person's mind. To start with, the system needs to be calibrated by having the subject watch a set of reference videos, so that the computer can figure out how to correlate the MRI patterns with the movies that are being watched. Once the computer is able to make this correlation, they then load up its database with over 18 million seconds of video chosen at random from YouTube.

Now they are ready to have the subject watch a different movie – one that wasn't in the original calibration set, or the YouTube sample – while the computer does its best to piece together its own reconstruction based on the fMRI feed from the cortex. It does this by pulling out any clips from the YouTube sample that seem to match up with the brain activity, and then it merges the top 100 matching clips to produce a best guess at what the subject is actually seeing. The resulting mind-movie is quite blurry but otherwise it bears a remarkable resemblance to the movie that the subject is watching, especially given the low spatial and temporal resolution of the fMRI images.[18]

Another UC Berkeley team carried out similar mind reading

experiments but with sounds instead of images. The volunteers in this case were being treated for epilepsy, so they already had electrodes embedded in their brains. Neuroscientists Brian Pasley and Robert Knight were able to use these embedded electrodes to observe the cortical activity that occurred as the subject listened to a series of spoken words. The brain signals were then sent to a computer which translated them into audible sounds.

"This study mainly focused on lower-level acoustic characteristics of speech," said Pasley. "But I think there's a lot more happening in these brain areas than acoustic analysis."[19]

For now, the computer only provides an approximate reiteration of the words that the subject hears, but Knight believes that it will eventually be possible to build a prosthetic device that can vocalize the words that someone is trying to say, by reading that person's thoughts.[20] Such a device could help individuals such as Stephen Hawking, the famous British physicist who lost the ability to speak as a result of a neurodegenerative disease.

Once we've gotten good at reading content from minds, it should be a fairly simple matter to create backups of the various sounds and images that are generated from these brain scans. Many futurists are hopeful that we will ultimately develop the technology to create much more detailed scans that will include all of the synaptic patterns representing an individual's mind. The most optimistic viewpoint is that this backup of the mind could be loaded into an artificial brain, whereupon it would be reanimated into a conscious identity.

This kind of speculation is a good example of the brain getting ahead of itself, and definitely takes us into Edelman's crackpot territory. It's in the realm of what my mother would call "far-fetched." In fact, British neurobiologist Steven Rose doubts that it will ever be possible to transfer a mind in this way, because of the complex ways in which the brain physically evolves over time in response to external stimuli.[21]

As long as consciousness remains mysterious, it is likely to hold up any plans we might have to make sentient machines. Nevertheless, there are many serious thinkers who believe that we are getting close to the day when machine intelligence takes over, an event that computer scientist and science fiction writer Vernor Vinge referred to as the singularity. Inventor and entrepreneur Ray Kurzweil is a leading proponent of the singularity idea, and he claims that our minds will soon be represented in software and people will merge with machines. We will be able to upload our minds into computers, and live within virtual worlds in a sort of digital version of heaven. The result will be an explosive expansion of intelligence which eventually consumes the universe.[22]

As we've observed, though, it's not a foregone conclusion that our minds can, in fact, be fully emulated in a computing machine. MIT robotics professor Rodney Brooks believes that consciousness can be replicated, although perhaps not in a computer.

> Although I do firmly believe that the brain is a machine, whether this machine is a computer is another question.[23]

In fairness to Kurzweil though, he doesn't seem set on digital computers. Rather, it seems he wants to drive home the point that machine consciousness is possible, regardless of what kind of machine you use. Even if it turns out to require a quantum computer, says Kurzweil, there are no insurmountable challenges from an engineering standpoint.[24]

This reference to quantum computing pays homage to British mathematician Roger Penrose, who believes that consciousness can only be explained through quantum mechanics. Koch, however, thinks it unlikely that any quantum effects are involved. The more we learn about the brain, the more it becomes apparent that our thought processes are all about neurotransmitters, ion channels and receptors – all involving fairly large scale interactions that are outside of the sub-atomic world where quantum effects are important. The creation of an action potential, says Koch, "is a large scale operation that will destroy any coherent underlying quantum state, if there would be quantum states inside the neuron."

He also says that "there is no evidence whatsoever for any quantum mechanical effect at the level of individual channels."[25]

Kurzweil believes "that there is a link between consciousness and quantum decoherence. That is, consciousness observing a quantum uncertainty causes quantum decoherence."[26]

Koch, however, is equally skeptical of this notion that quantum decoherence is somehow coupled to conscious observation. There is no logical reason to bring quantum mechanics into the mix, he says, other than the dubious assumption that since both consciousness and quantum mechanics are mysterious, they must somehow be related.[27]

Kurzweil doesn't seem to care how we build machine consciousness, as long as it gets built. He's confident that one way or another, consciousness will emerge in machines, and he predicts that the singularity will happen by the year 2045. That's an ambitious target, to say the least. It's true that humanity has met some ambitious targets in the past, including the one set by President Kennedy in 1961 when he committed the nation to "achieving the goal, before [the] decade [was] out, of landing a man on the Moon and returning him safely to the earth."[28] Kennedy knew it wouldn't be easy.

> But if I were to say, my fellow citizens, that we shall send to the Moon, 240,000 miles away from the control station in Houston, a giant rocket more than 300 feet tall, the length of this football field, made of new metal alloys, some of which have not yet been invented, capable of standing heat and stresses several times more than have ever been experienced, fitted together with a precision better than the finest watch, carrying all the equipment needed for propulsion, guidance, control, communications, food and survival, on an untried mission, to an unknown celestial body, and then return it safely to earth, re-entering the atmosphere at speeds of over 25,000 miles per hour, causing heat about half that of the temperature of the Sun . . . and do all this, and do it right, and do it first before this decade is out—then we must be bold.[29]

Yet, in many ways, going to Moon was relatively easy compared with the task of figuring out consciousness. For the most part, the Moon challenge was more of problem for engineers and project managers than for scientists. Despite the difficulties that lay ahead, Kennedy and his advisors knew that if they were to beat the Russians to the Moon, it was mostly a matter of marshaling the country's resources, and providing the necessary leadership. Building the infrastructure for space travel called for the orchestration of 400,000 souls, all working toward a common purpose. Space suit assembly, for example, required the skills of five hundred seamstresses. In addition to the aforementioned giant rocket, we needed a space port with a massive vehicle assembly building, which by itself was a huge undertaking.

Yet, while the concept of a Moon landing was unfathomable to my great grandmother, there were many sharp minds in the 1960s who knew what was needed to achieve that goal. It was also fairly easy to define the criteria for success. Once you get a man to set foot upon the Moon, and bring him safely back to Earth – you're done. Consciousness, however, is a different matter, and not so much an engineering challenge that can be solved with lots of dollars and resources. After decades of research, we are still asking some fairly fundamental scientific questions about the human mind, such as "what is it, exactly, and how is it produced?" In that sense, the path to understanding consciousness may be less like a Moon shot, but perhaps more like the quest to understand genetics.

The study of genetics garnered considerable excitement in 1953, when James Watson and Francis Crick walked into the Eagle pub in Cambridge and announced that they had discovered the "secret of life." But even with all this excitement, it still took another fifty years for scientists to come up with a complete map of the human genome, nature's blueprint for humanity. There were high expectations that once this map was in place, revolutionary changes in medicine would follow, and although this is by no means overstated, the revolution is taking longer than many had expected.

Entering the nucleus of one of your cells is a bit like walking into an auto parts store, although I'll admit, not exactly. To start with, the nucleus doesn't have rows of shelves with little boxes containing the various proteins you're looking for. Instead, the proteins are made to order, using the recipes found in your DNA. When a cell needs some new auto parts, it needs to cook them up in its nucleus kitchen. In this half-baked analogy, the genes are the recipes, and the proteins are the various auto parts. When scientists announced that the human genome project was complete, what they meant was that we now have a complete recipe book, obviously an extremely valuable tool for biologists. For example, having this recipe book makes it much easier for researchers to identify the genes that may be implicated in diseases. They can start by looking through the table of contents to make a guess at which group of recipes might have gotten messed up, so that they can zero in on the likely culprits. This kind of research used to take a decade or more, whereas now it can be completed in months.

The recipe book doesn't solve all of our problems, however. Even when I exit the auto parts store with the correct part for my make and model, which happens about fifty percent of the time, I still often have difficulty figuring out how to install the darn thing correctly. I'm remembering a certain headlight bulb that refused to fit into its receptacle, because it had the wrong shape. The same applies to proteins; they won't work if they have the wrong shape. Each protein interaction can trigger a series of additional biochemical events, in a sort of domino effect, and these chain reaction pathways are where things really get complicated. Proteins are also involved in regulating gene expression, which means that they can go back to the recipe book and order up some additional proteins.

One interesting finding from the Human Genome Project was that there are fewer protein-coding genes than were originally anticipated. There are only about 25,000 protein recipes in the book, which seems a little low when you consider that there are 30,000 parts in a typical Toyota. This is a little misleading though, since I'm implying that each gene translates to exactly one protein. In fact, there are many more proteins than genes, so as long as we're talking about the total number of parts, we can at least say that the human body is more like a space shuttle than a Toyota.

The true level of complexity, of course, depends upon how the parts get used. You can make a lot of interesting meals with 25,000 recipes. Or, to use yet another analogy, I am continually impressed by the seemingly endless procession of LEGO creations that emerge from my five year old son's bedroom on a daily basis. These creations, or product ideas as he likes to call them, are all made from the same bucket of LEGO bricks, yet each new construction is unique. Likewise, we often hear that humans have a lot in common with chimpanzees; that is to say, for the most part we share the same bucket of bricks. How then can we explain the rather obvious anatomical differences between species, if they have very similar sets of genes? The answer, it seems, is that much of the diversity that we see across the animal kingdom is caused, not just by the genes themselves, but by how these genes are regulated.

To get an idea of how this regulation works, consider that in addition to the protein recipes, we also have a lot of DNA that doesn't code for proteins, which is referred to, naturally enough, as non-coding DNA. Some pieces of non-coding DNA play an important role in allowing genes to be switched on or off. An important point to note is that you can have several switches that control the same gene. I have a similar situation in my house, where the light at the top of the stairs can be controlled by three different switches. If one of the switches is disabled, we can still operate the light from one of the other two locations. The same principle was adopted by nature in the evolution of organisms.

The standard account of evolution is that once in a while a favorable mutation comes along that somehow embellishes an organism's love making skills, thereby leaving more progeny around to propagate the genetic improvements to future generations. But instead of changing the genes, another option is to install a new switch, allowing an existing protein recipe to be used in a different part of the body. Other evolutionary changes might result in a switch being disabled, thereby shutting down that particular protein pathway, but without preventing the same protein from being produced elsewhere via a different set of switches. So, to get a better understanding of what makes us tick, we need to keep track of these switches, along with where and when they get used. That means we may need a more detailed cross reference section in the back of our recipe book.

To make things even more interesting, there is another kind of control mechanism involving what are called non-coding RNA genes. In the normal process of protein manufacture, a gene gets transcribed via a molecule known as RNA, which is then used to manufacture a protein. However, there are times when RNA plays a different role, such as regulating the expression of a gene. In this case, the RNA is essentially acting as another kind of switch, usually resulting in the suppression of certain recipes.

Almost sixty years have passed since Crick bragged about having found the "secret of life," and yet it's pretty clear that life hasn't given up all its secrets. Now that we can understand what life is, in terms of mechanistic, molecular steps, you'd think that we could enter all this data into a computer and have it predict which proteins will be synthesized, along with their shapes and functions. Yet even when you know the shapes of proteins, it can be very difficult to predict how they will behave in the complex environment of a cell. Some problems are difficult to solve.

We had high hopes when President Nixon declared war on cancer in 1971, but forty years later, that war is still being waged. For decades we've talked about developing ways to more effectively target cancer cells, and we've imagined using specially engineered viruses to eliminate diseases by correcting genetic mistakes. But the concept of gene therapy took a serious blow in 1999 with the tragic death of Jesse Gelsinger. The eighteen year old suffered from a particular kind of liver disease known as OTCD, and researchers at the University of Pennsylvania were experimenting with engineered adenoviruses to try to come up with a cure. Gelsinger was recruited to help evaluate the safety of the new technique, but unfortunately his body reacted badly to the modified virus; his immune system was thrown into shock and within a few days he was dead. The incident still serves as a reminder of why it is important to be meticulously slow when developing new medical technology.[30] [31] There's little doubt that miracle cures are still in the offing as we continue to perfect these techniques, but there remains a lot of work to be done; probably enough to keep our biochemists off the streets for the foreseeable future.

In later years, Crick turned his attention to the problem of consciousness, perhaps hoping that he might one day burst into the Eagle and announce that he had figured this one out as well. But although he made a huge contribution to the subject, Crick could never claim to have cracked the case, and he can't have been feeling too optimistic when he admitted to Searle that "you know, consciousness – that's a lot harder than DNA."[32]

The topic of artificial intelligence (AI) is another in which expectations have greatly outpaced results. Broadly speaking, AI is about building computers and robots that can do the kinds of things that we humans take for granted, a subject that has seen slow and steady progress, with the emphasis on slow. Years have been spent just trying to build machines that can move around. We now have robots that can walk, and some that are getting better at autonomous navigation, but as science writer Susan Hassler notes, "This is hard scientific and technical work."[33]

We've also had some success with so-called thinking machines. In 1997, chess champion Gary Kasparov was beaten by an IBM computer known as Deep Blue, and in 2011, another IBM machine named Watson went up against two former champions in the popular TV quiz show, *Jeopardy*, and walked away (though not literally) with the million dollar prize. IBM donated the money to charity.

To be able to out-think a human mind, Watson was supplied with four terabytes of information, including the entire text of Wikipedia, along with the remarkable ability to learn the relatedness of different facts. But even though it's been touted as the smartest machine on Earth, Watson still doesn't have a mind, and its creators admit that it will be long time before they can build a machine with anything like the cognitive powers of the human brain.

On the other hand, it would be wrong to assume that all AI specialists are hell-bent on replicating consciousness. One can easily think of a variety of uses for a Watson-like device, that don't require it to be conscious. A medical version, for example, might be used by doctors to check their diagnoses, or to see if there is anything in the latest research that might be relevant to their patients. In fact, non-sentient machines might always be the preferred approach when developing AI for commercial, scientific or military applications.

Sentience aside, we will, nevertheless, design our machines with more human like qualities, to make it easier for us to interact with them. That's the view of Cynthia Breazeal from MIT, and the reason behind her attempts to endow robots with facial expressions which simulate human behavior and emotional responses.

In the coming decades, we can look forward to getting acquainted with more computing machines like Watson, with specialized, super human intelligence, narrowly focused on specific kinds of tasks. While some AI aficionados remain optimistic that consciousness might show up in one of these machines as an emergent quality, most people, including Kurzweil, are now of the opinion that the only way to build a machine that's actually conscious, is via some kind of reverse engineering of the brain. That's what neuroscientist Sebastian Seung of MIT has in mind.

Now that we have mapped the genome, Seung thinks it's time to get started on what he calls the connectome, a complete picture of the neural wiring in the human brain.

Searle, for one, is gratified that AI experts are finally beginning to realize that if you want to understand the mind, you can't ignore the brain and just hope that computers will be sufficient. Kurzweil predicts that conscious machines will quickly surpass human intelligence, but in many ways our unconscious computers have already done that. If we ever get around to building something that's conscious, we'll be looking for social intelligence and not the kind that can outmaneuver us at Jeopardy. To start with, it would be quite an accomplishment to build a machine with the social awareness of Searle's dog, let alone the cognitive skills of a two year old child. That would be a far more impressive achievement than the creation of intelligence per se.

Perhaps we should pause here for a moment, and consider why we might want to build a conscious machine in the first place. This may seem like a silly question, given that it's what science fiction writers and AI enthusiasts have been dreaming about for years. We've all become accustomed to the idea that we will eventually have robot companions that are more than just mindless automatons, and for engineers, there would certainly be a huge novelty factor in being able to build machines that can really think. We could treat such machines as pets, I suppose, but then as I keep telling my daughter, a pet is big responsibility.

Realistically, since it now looks like any progress in that direction will inevitably be driven by the quest to understand the brain, it seems to me that any attempts to start replicating what brains can do, will be undertaken at least initially for medical reasons. Understanding the brain is primarily a job for neuroscientists rather than AI specialists, and from a human perspective the main focus will be to develop the means to help those who are suffering due to debilitating cognitive disorders. Prosopagnosia, for example, is one such disorder that prevents people from recognizing faces, including the faces of family members. All we know so far is that our ability to quickly distinguish one face from another seems to involve an area of the brain called the fusiform face area. Presumably then, some instances of prosopagnosia may be the result of damage or disruption to that part of the brain. Once neuroscientists identify the underlying causes of such problems, future surgeons may then be able to effect repairs. Perhaps they will have a way to regenerate damaged tissue, or rewire the neurons, or maybe they'll splice in some kind of prosthesis, like the artificial hippocampus being developed by Theodore Berger of the University of Southern California.

In April 2011, Alice Parker, also at USC, announced that her team had successfully assembled an artificial synaptic circuit using neurons made from carbon nanotubes. Decades from now, Parker believes we may be able to build a network of these synthetic neurons to replace a damaged part of the brain, or perhaps even construct an entire synthetic brain. She also thinks that we'll see the same technology making its way into AI as well, in the form of auto navigation and safety systems. Parker's big challenge in near term is to find a way to use these nanotube circuits to implement plasticity, the structural changes that occur in the brain to accommodate learning and memory.[34]

If we ever manage to build a brain of the kind that Parker has in mind, what would we use it for? Would we make lots of them, and then train them to be our slaves? Or might we be looking at brain transplants? Having your brain replaced with an artificial contrivance is bound to be more complicated than getting a new heart, or a kidney. While the last few decades have seen many successful heart transplants, it's still by no means easy. The first challenge is to find a suitable match and to make sure that the donor's heart doesn't get rejected by the host body. When a normal transplant operation is not possible, some patients now have the option to have the failing heart replaced with an artificial device. Artificial hearts are now a reality, though they still have a long way to go before they can match the performance of a human heart, and are usually regarded as a measure of last resort. In addition to being able to pump blood around the body, manufacturers need to ensure that their inventions can continue to function reliably for many years, if not decades, and that they don't introduce other complications, such as blood clots that could lead to strokes.

Then there's the problem of how to power the device. Atomic power seemed to work well for Steve Austin, but that's not really a viable option in the real world. SynCardia's CardioWest is a pneumatic heart, operated via air pressure that's fed from an external source through a couple of tubes that protrude from the patient's abdomen. Abiomed takes a different approach with their solution, called AbioCor, which is fully enclosed within the patient's body. AbioCor uses an implantable battery pack that is recharged by placing a magnetic contact against the skin, obviating the need for any protruding wires or tubes and thus reducing the risk of infection.

It may be, however, that long before these artificial organs have been perfected, we will have already figured out how to grow replacement organs with organic material, instead of relying on electrical pumps and batteries. Researchers are already using new kinds of scaffolds to coax human cells to grow into new lungs, hearts and kidneys in the lab. With this approach, transplant patients will no longer need immunosuppressive therapies because the new organs are fabricated using their own cells. Most importantly, it could mean that patients will no longer die while waiting for replacement organs.

One way or another, if we ever decide to start transplanting brains, we'll have a whole new set of problems to worry about. To begin with, a brain is unquestionably a more complicated artifact to install. Brains not only have to be connected up to arteries, but there's also a lot of intricate connectivity to the rest of the nervous system as well. Artificial hearts are relatively simple by comparison, and are basically just sewn into place. They even come with what are called "quick connects" that make it easier for surgeons to attach them to the blood vessels.

Of course the whole business of transplanting brains makes no sense at all, unless we also have some way to preserve the content of the old brain, before it gets thrown away. Otherwise, well, you've lost your mind, haven't you? As I see it, there are at least two sides to the problem. First of all, we need to be able to extract information from the human brain – to read the patient's mind, as it were. All of the memories residing in the brain must be captured using some kind of scanning device, preferably one that's non-invasive. Secondly, we'll need the technology to load this information into the new brain. Needless to say, neither of these things is currently possible, nor is there a reasonable consensus on whether any of this is even plausible. But let's explore the concept a bit more.

Off the bat, one would assume that for brain transplants, you're not going to find too many donors lining up, for the obvious reason that if the brain is still viable, the donor will probably want to hang onto it. In any event, I would imagine that having your mind forced into an already existing brain would be out of the question, since this would involve erasing someone else's mind; a move that I expect would be frowned upon. Also, since the actual structure of a brain reflects the memories and experiences of its owner, we would need to physically rewire a brain in order to literally change someone's mind into that of another person. Apart from the fact that this is ethically disturbing, it might also be very difficult to do. It may be easier all around to custom build a new brain for each patient.

In the normal course of brain development, neural pathways are formed in response to what the brain experiences. Presumably if you took two brains, and exposed them both to exactly the same external stimuli, you'd eventually end up with two identical minds. In reality though, we never see this happen, since each brain has a different genetic and biochemical basis to begin with, and since no two brains are ever exposed to exactly the same external stimuli.

Rose notes that when he looks down the street and observes an approaching bus, his mind interprets the scene in a way that depends, not just upon the sensory input, but also upon the predisposition of his brain. Apart from the genetic influence, there are lots of other factors in his personal makeup that affect how he perceives the world. He tells us that, in order to fully appreciate his point of view, our mind-reading device would need "to have been coupled up to [his] brain and body from conception – or at least from birth – so as to be able to record [his] entire neural and hormonal life history."[35] However, even if Rose is correct, and his innermost perceptions are to remain enigmatic, that doesn't mean that we couldn't, in theory, replicate his enigmatic self by making an exact copy of his brain.

Rose's thought experiment reminds me of an old *Star Trek* episode in which Lieutenant Uhura's mind is completely erased by a less than empathic alien probe. Dr. McCoy is unfazed, and assures the Captain that he can restore Uhura's mind, by reeducating her brain using the ship's library tapes. I'm thinking that this sounds like an oversimplification on the good doctor's part. Unless those tapes include a complete dump of the lieutenant's synaptic patterns, and McCoy has the wherewithal to use this data to fully restore Uhura's neural pathways, I'd say there's a fair probability that they just lost their communications officer. A subsequent scene shows Uhura learning to read "The dog has a ball," so it's not looking good.[36]

A slightly more believable arrangement might entail directly loading the patient's memories into the replacement brain as it is being synthesized. Rather than feeding this data through the normal sensory channels, imagine instead a data transfer process involving more precise manipulation of the neural wiring, so as to replicate the patterns from the source brain. If we use a synthetic device, we might even find a better way to implement plasticity, other than having to physically change the neural structure. This could make it easier to insert a new mind, assuming of course, that the implementation results in the artificial brain having the same causal properties as a real brain.

It still doesn't sound particularly easy to do, but on the other hand, neither does it sound like the kind of thing that will remain impossible forever. We are, after all, looking to replicate something that actually exists, namely a human mind, as opposed to something that might never exist, such as warp drive. Rose is correct in pointing out that his memories were formed through a very convoluted process, but the fact remains that these memories are nonetheless stored somewhere inside his head. The data exists, so it's hard to argue that engineers of the future might not be able to access the contents of our heads using more advanced scanning techniques. Now that Gallant and Knight have started down that road, I'd imagine there's no turning back.

Once we've restored the patient's mind into its newly minted brain, the next step would be to install the new brain back into the original body, and then carefully connect it to the various peripheral systems. Alternatively, it could be loaded into an android body, assuming the technology were available. One advantage of this route would be that you could dispense with all of the organic parts, replacing them with more durable components that come with extended warranties. The concept was dealt with in yet another 1960s *Star Trek* episode, one that was first shown when I was exactly one year old. This time, Lieutenant Uhura, her mind now fully restored, is offered the chance to become immortal. The story takes place on a planet inhabited by tech-savvy androids that have the ability, we are told, "to place a human brain within a structurally compatible android body."[37] However, it wasn't at all clear how this procedure would provide Uhura with immortality. Her android body might not wear out, but what about her organic brain?

Another alternative would be to completely avoid the hassle of having to install the new brain into a physical body, either human or android. We'd need to connect it to something, of course, since the only way in which a brain can interact with the outside world is via chemical and electrical signaling. But what if that signaling was orchestrated by an artificial intelligence that could replicate the kinds of patterns the brain would expect from the real world? We could, in other words, connect the brain to a sophisticated computer system, so that instead of a real body, you'd have a virtual body that could inhabit a computer simulation of reality. If the simulation was accurate enough, this virtual existence might feel very much like the real thing.

What if we manage to grow a synthetic brain before we develop the technology to load it up with synaptic patterns from a real brain? In this case, a brain transplant would be out of the question, although we could presumably allow the synthetic brain to grow into a new sentient being, if we wanted to do that.

Before we start capturing minds and transferring them from place to place, we're going to need some kind of advanced brain scanner – one that has "not yet been invented," as President Kennedy would say. More than likely, researchers will need these tools to help them to properly identify the neuronal correlates of consciousness, and to come up with a theory that explains exactly how consciousness occurs. That means we may need to build the scanners first before we complete work on the synthetic brains. An optimistic goal would be to develop a high resolution scanner that could capture all of the neuronal connections, synaptic strengths, and firing patterns that represent the conscious mind, and then we'd have the ability to copy and record all of this data, even if we didn't yet know how to reanimate the mind by restoring the data back into a functioning brain. Some future entrepreneur will no doubt develop a business plan around giving people the opportunity to have their minds backed up to a secure storage system, just as some people today opt to have their heads frozen when they die, in the hope that some future generation might revive them. Given the choice, I reckon I'd prefer a brain scan to a brain freeze, assuming of course that it was a non-invasive scan. I would think that the data storage cost for a mind would be a lot less expensive than the cost of maintaining a frozen head in liquid nitrogen. Another advantage of the brain scan approach is that it doesn't involve you having to actually die.

When we finally get an artificial brain up and running, Kurzweil says that it will very quickly become super intelligent. These machines will then manufacture even more advanced machines, and the resulting exponential growth in intelligence will be what determines the course of humankind. Is this a realistic scenario? If we're looking for problem-solving ability, then we must acknowledge that there are already plenty of machines that can surpass human intelligence.

"Any pocket calculator can beat any human mathematician at arithmetic," reminds Searle.

"It is perfectly legitimate to say that the chess-playing machine has more intelligence, because it can produce better results. And the same can be said for the pocket calculator."[38]

That is to say, we've successfully managed to create a lot of artificial intelligence, without the need for sentience or consciousness. As Koch explains, consciousness is a whole different ball game.

> There's no direct relationship between intelligence and consciousness.[39]

Consciousness is a biological phenomenon, and so it needs to be appreciated in a biological context. It's worth reiterating that any progress toward a theory of consciousness, or any attempts to replicate consciousness artificially, will require a greater familiarity with brains rather than computers. In particular, our exploration of the brain is quite a profound undertaking, as it deals head-on with the broader question of what it means to be human. As such, it's a subject that more properly resides in the human domain, rather than in the applied fields of computer science and AI. That means that the qualities that eventually get replicated in our sentient machines will be human qualities, and the minds that those machines will support, will be human minds.

So as we wonder about the nature of artificial brains of the future, it's worth remembering that we'll still be dealing with human nature. If we end up getting our minds placed into external devices, then we're likely to preserve that which is important to us, and in my mind, super intelligence doesn't top the list. It depends, of course, upon how much we value intelligence, by which I mean the ability to solve problems efficiently. Our brains evolved that we might have the skills to interact with each other in larger groups. Thus we are primarily social animals, which is why social awareness is generally seen as more important than an ability to solve differential equations. Our kids may not always show the same level of intelligence as adults, but we're still pretty fond of them nonetheless. As far as Ackerman is concerned, intelligence is not necessarily the key to a better life.

> Studies show that the IQ range of most creative people is surprisingly narrow, around 120 to 130. Higher IQs can perform certain kinds of tasks better – logic, feats of memory, and so on. But if the IQ is much higher or lower than that, the window of creativity closes.[40]

It seems more likely that we will continue to use machines to solve problems for us, while we concentrate on the important stuff. The prospect of cognitive biomimetic, prosthetic devices or artificial brains lets us imagine a future where people are more closely integrated with machines, but this integration is already underway to an extent. Wherever I happen to be, I can now tap into a planet-wide information network, and can answer any question that's put to me, just by talking into my smartphone. One of my colleagues recently demonstrated that he could do this even after a few beers. We can think of the smartphone as an extension of human intelligence, and maybe in the future people will decide to have smartphone equivalents connected directly to their brains, so that they'll always have them at the ready.

But has this added intelligence become the driving force in society? Have we placed a premium on intelligence for its own sake? I don't think so. Instead, as our devices become more intelligent, it appears that we are just using these devices to find ever more interesting ways to do what we've always done, which is to engage in social interaction. It might not sound as dramatic as Kurzweil's super intelligence scenarios, but that's essentially what people are about. We may agree with Carl Sagan when he says "It is better to be smart than to be stupid,"[41] but on the other hand if intelligence confers such a huge advantage, then why aren't there more super intelligent people? As I recall, we had a super intelligent person in our twelfth grade class at school, but if I remember correctly the consensus was that he was also quite unpleasant.

What about the possibility, ever present in science fiction, that a malicious, sentient machine will someday decide take over the world? I guess there's a non-zero risk that this might actually happen, but then we have to ask ourselves whether such an attempt would succeed, or if the machine would just end up making a nuisance of itself. In all probability, it would be the latter, in which case we would undoubtedly take measures to ensure that it never happens again. I'm suggesting, in other words, that the process by which artificial brains evolve will most likely result in the preservation of human values. Sentient machines of the future will therefore have a lot in common with the sentient machines of today, regardless of whether they have organic brains. They will reason things out within a human framework, because, in essence, they will be human.

6

⌘

Wishful Thinking

One of the things that Granddad Waine told me was that when we go to heaven, we become super intelligent. We get to share in the glory and wisdom of God, and we will know everything there is to know. I suppose this is like a spiritual version of what futurist Ray Kurzweil predicts for machine intelligence in the near future. Anyway, as a kid, I remember wondering what it would be like to know everything. It occurred to me that most of the time we're not actually thinking about all the things that we know. If I know something, it just means that the information is stored somewhere in my head, so that it can be accessed at the appropriate time. If I knew everything, presumably that would mean that I'd have access to a much bigger storehouse of information. In my case, I can't always be sure that the knowledge will be accessible when I need it. For example, in my college days, I used to know a lot more about nuclear physics than I do at present. In trying to recall the facts, I realize that I have forgotten most of them, maybe as a result of that LTD process we talked about earlier. Perhaps this will be less of a problem as we increasingly rely on machines to remember things for us.

It's a generally held maxim that the more knowledge you acquire, the more mature you become, or at least that's how it's supposed to work. The worldly wise are usually better informed than the rest of the community. We expect that such knowledgeable people will be less biased in their decision making, and we are more inclined to trust their judgment. If knowledge leads to maturity in individuals, then perhaps the same can be said for society in general.

On a visit to the California State Railroad Museum in Sacramento, we were accompanied by a well-informed tour guide who told us all about the completion of the transcontinental line in 1869. Workers on the Central Pacific section started out in Sacramento, California, and eventually met up with the Union Pacific team that worked its way westward from Omaha, Nebraska. The two teams came together at Promontory Summit, Utah, where the final "golden" spike was driven into place.

On the western side, much of the work was done by Chinese immigrants, who faced the especially difficult challenge of having to blast their way through the Sierras at an elevation of seven thousand feet. Many lost their lives due to the various hazards of the job. However, when you look at the commemorative photograph on display at the museum, there are no Chinese faces to be seen. Apparently they were deliberately excluded from the photo, despite their huge contribution to the project. "Thankfully," said the tour guide, "we treat people a lot better than we did in those days."

It's true, isn't it? As we remember how willingly the world once embraced the shameful practice of slavery, it is both remarkable and gratifying to live in an age that no longer tolerates blatant discrimination. In the last century, we've seen a multitude of societal reforms that have strengthened the civil rights of minorities, challenged the unfair treatment of women, and forged new health and safety regulations to protect the working population and the environment. Society has matured, and some of the changes have been surprising.

I'm thinking back to those cold Irish winters, and the miserable rainy days when I would take the bus home from school. They were double-decker buses, just like the ones in London, and on days when it was wet there was no standing room downstairs, so the conductor would make us go up top where we were greeted by a dense cloud of cigarette smoke. Smoking was only permitted on the upper deck. You couldn't see outside, because the windows were always closed and drenched with condensation. On one occasion, as I struggled to breathe in the blue fog, I tried opening one of the small windows, only to have it slammed shut again by an old lady who had no appreciation for fresh air. I never expected that, within a few decades, smoking would be banned in all public places, including bars and restaurants. Whoever had heard of a smoke-free Irish pub?

So what's the reason for all this change? In the case of the smoking ban, it's most likely because many more people are now aware that smoking can kill. So much is now known about the detrimental effects of cigarettes, that the idea of allowing smoking in the workplace has become untenable. I was sold on the danger of second-hand smoke after seeing an interview with Roy Castle, the famous British entertainer. He was best known as a trumpet player and after years of belting out music in smoke filled bars and clubs, his lungs finally succumbed to cancer. These kinds of stories have the potential to shift public opinion, and over time they can lead to progressive changes in society.

Figure 7. East meets west - The completion of the transcontinental railroad at Promontory Summit, Utah.[1]

In 1911, for example, the plight of sweatshop laborers became an important issue after the infamous Triangle factory fire in New York. When people read about the tragic deaths of so many young women, it had such a heart wrenching effect on the public conscience that the resulting investigation spurred ground breaking legislation to reduce the risk of similar occurrences. In the second half of the century, television had an even more potent effect on public perception. We can see how much things have changed from recordings of old television broadcasts. Coverage of the Moon landings, for example, doesn't show any women in the mission control room, and the men don't seem to have any qualms about smoking near high tech equipment.

If seeing is believing, then television has surely changed the way we think about the human condition. We can use it to travel back in time, to see what life was like on the slave ships, or to relive the civil war. When my daughter was six, she told me that it would be wrong to deprive her of television, because it's her window to the world. Back in the 1980s, we could either watch broadcast TV, or we could schlep down to the video store to rent a movie. Today, through the wonders of information technology, I have instant access to any TV show or movie that interests me. I can also access online books and remote databases, attend conferences, review scientific journals, and get satellite imagery and video feeds from all over the globe. And the nice part is that I can do all of this without leaving the house. All of this content contributes to a more enlightened and mature society. We no longer have to turn to the village elder, or the seasoned traveler, to get a deeper perspective on life.

As well as having more content, and more ways to access it, information technology also provides new communication and social networking tools that make it easier for individuals to collaborate, and to participate in both local and global communities. In 2011, for example, it was reported that social networking technology played a key role in the reformist movement to overthrow Egyptian dictator, Hosni Mubarak. This is the essence of intelligence – having the ability to accumulate facts about the world, and then apply this knowledge effectively to achieve the desired outcome. Now that we are developing the tools to do this on a planetary scale, perhaps we're witnessing the emergence of a global intelligence, which could lead to a more democratized and socially responsible global community. As parents, we know that knowledge and education are the means for our kids to become well-rounded individuals. In the same way, our public libraries and information technology framework provide the foundation for a well-rounded civilization and an enlightened populace. In the long run, these are values that will ensure our survival.

We do not yet have artificial brains, but that doesn't mean we've been limited to the confines of our heads. Language allows us to pool our minds and to work together as a collective. We've discovered new ways to share information, and to make it available from one generation to the next. These are the basic skills underlying the evolution of human culture, and each new development adds to the rich tapestry of our existence. The more we know, the better we become. If we eventually decide to upgrade our brain hardware, no doubt it will mean that we'll be able to think faster, and maybe we'll also find it easier to communicate and to tap into our shared storehouse of information, instead of evolving into obsessive, super-intelligent machines, bent on world domination. Perhaps the transition will have a more mellowing effect on our personalities, and with our newly sharpened minds we'll all become well-informed, benevolent machines, with a positive outlook on life.

Some have suggested that our new brains will generate such an appetite for novelty that we'll need to start building giant computers to host the new virtual worlds which we will inhabit. Kurzweil is confident that as soon as we've harnessed the computational capacity of the entire planet, we'll then be compelled to ravage the rest of the universe, in a frenzied quest to expand our intelligence.[2][3]

Of course, it's possible that things may take a different turn. For one thing, ravaging the universe may be easier said than done. If it's inevitable, then shouldn't our telescopes have furnished some indication of such ravaging behavior elsewhere in the galaxy? Furthermore, a lot will depend upon the nature of the intelligent machines, and whether or not they are motivated to start marauding through the universe in the first place. If, as I've speculated, we're still talking about human nature, then it's possible that they might have a more conservative mindset. Remember LTD? It's the abbreviation for long term synaptic depression; the process that some neuroscientists believe may be responsible for clearing out old memory traces in the brain. If the goal in creating an artificial brain is to replicate the biological activities that establish a human mind, then this must include the neural mechanisms that allow us to forget. Otherwise, if we modify the properties of the brain in an attempt to prevent memories from fading, we risk compromising the integrity of the mind we are trying to preserve.

The fact that we forget things is what makes life interesting. Novelty has its attraction, for sure, but many of the sweeter aspects of existence involve being able to relive the past. I like to revisit my favorite restaurants, and enjoy having meals with old friends. I like to travel to familiar places, and to be surprised by the reappearance of old memories. We have a special affinity for sameness. Every Christmas Eve, I open a bottle of wine and sit down to watch Albert Finney's version of *Scrooge*, even though I've seen the same movie many times over. I'm not suggesting that we won't need a decent computer to recreate all of this, but I might be willing to settle for something a little smaller than planet-sized.

Hopefully then, the increased intelligence of artificial brains will result in a society that's even better at living up to its moral responsibilities, with expanded opportunities for human creativity and self-fulfillment. Instead of trying to kill everyone, perhaps these machines will be more inclined to show a little kindness and respect. After all, what would motivate an artificial brain to turn malevolent? For that matter, why does it happen in regular mortals? It's a question that news anchors like to ask themselves every time we have a mass shooting incident. Neuroscientists Kent Kiehl and Joshua Buckholtz point to the paralimbic areas of the brain, noting that psychopathic tendencies seem to arise when these areas are damaged or undeveloped.[4] Psychopaths are known for their inability to feel empathy for others, a trait that neuroeconomist Paul Zak suggests may also have something to do with how their brains use a particular neuromodulator called oxytocin.[i]

Zak devised an experiment to measure trust, and to see if this trait had any ties to oxytocin. The experiment involved a game of trust, in which subjects were each given a sum of money, and then had the choice to give some of that money to their partners in the game. For example, suppose subjects Susan and Bob each receive $10. Susan then chooses to give $6 to Bob. The $6 is tripled, so Bob now has 28 bucks. He can decide to give some of the money back to Susan, although the amount returned is not tripled.

The basic premise is that Susan gives money to Bob because she trusts him to pay her back. What Zak found was that whenever subjects decide to part with their money, the recipients feel trusted and thus more oxytocin shows up in their blood. The more Susan decides to give, the

---

[i] A neuromodulator is a neurotransmitter, but instead of being used at a single synapse to allow one neuron to talk to another, neuromodulators, once released, can diffuse through the brain and affect other, more distant neurons.

more Bob feels trusted, which means his brain produces even more oxytocin. When oxytocin levels were artificially boosted via a nasal spray, there was a noticeable increase in trust levels – in other words, the subjects gave away more money.

The experiments seem to show that this particular brain chemical can produce feelings of trust as well as an inclination to be trustworthy. A small number of individuals returned little or no money, and scored very poorly on the trustworthiness scale. Their oxytocin levels were abnormally high, which suggests to Zak that there might be something wrong with their oxytocin receptors. "Tellingly," he says, "the highly untrustworthy possess personality traits that resemble those of sociopaths, who are indifferent to or even stimulated by another's suffering."[5]

Suppose someday we can be more precise about the neurological deficiencies that cause individuals to have psychopathic predilections. Would it be acceptable to correct these deficiencies, or might such an attempt be regarded as an unacceptable breach of privacy? It could be argued that once we start fixing defective brains, we run the risk of trying to engineer the perfect mind, which would surely be the ultimate assault on personal freedom. On the other hand, what if we could help people with prosopagnosia, and restore their ability to recognize faces? Is that any different than being able to fix the brains of psychopaths so that they can feel empathy for others?

Perhaps it would depend on whether the surgery was elective, or if was part of a mandatory government program to rehabilitate offenders. I expect we would want to avoid anything that might present the specter of social engineering through medicine. That poses the question of whether we would replicate malevolent tendencies in our synthetic brains of the future. If we're going to have psychopaths on the loose, it might as well be in a virtual world where at least they won't be able to kill anyone, although that's not to say they wouldn't find some other way of getting into trouble. In any case, it seems a little premature for us to worry. Perhaps our engineers of the future will find it easy to identify and repair the personalities that are obviously flawed without interfering with the rest of us, who are just different.

While it is fun for science fiction writers to fantasize about creating sentient beings like Lieutenant Commander Data, under what circumstances would we allow someone to deliberately set out to induce human-like consciousness in a brain-like device? Could such an entity be treated as a tool for science, or used for commercial purposes? If we truly manage to create an artificial brain that works just like a real human brain, then what we are describing is something that has the potential to become an actual person, but only if it gets the same care and nurturing that would be given to a human child. There's plenty of science fiction that deals with the ethical dilemmas and horrific situations that might come to pass if we start creating people in this way. Would we allow the creation of a disembodied mind, for example; an artificial brain living in a jar on a laboratory shelf, perhaps hooked up to a computer so that technicians could interact with it?

As long as the discussion is about human sentience, then the ethics of the matter are pretty clear. Human rights must apply to all persons, regardless of what materials their brains are made out of. If we were to allow experimentation on sentient, artificial brains, we might just as well allow people to experiment on the brains of children. In an interview with Science Magazine, Henry Markram says that he doesn't know if his human brain simulation will ever become conscious or not.[6] I'd say the chances are it won't, but if Markram does end up building something with the causal powers to create consciousness, then someone will need to think seriously about the implications of proceeding in that direction.

Of course, other animals may be conscious too. Neuroscientist Christof Koch admits that even bugs may have some form of consciousness, since "We have no idea what the minimal complexity is of a brain necessary for there to be consciousness."[7]

Koch tells us that this realization led him to become a vegetarian, and it's also the reason he no longer kills bugs. Some might suggest that one good reason for intentionally creating new conscious entities in the lab would be to reduce the need to experiment on animals. Even so, one would hope that any such experimentation involving consciousness would at least be done for medical science, and not for developing new game consoles, for example.

Since the study of consciousness is closely linked to medical research, our near term goal will be to understand the processes underlying specific cognitive functions, so that we can find out how to restore this functioning in brains that have been damaged or impaired by disease. As we've speculated, this might involve the use of artificial prostheses, similar to the electronic chip that Theodore Berger's USC team is developing to take the place of the hippocampus; the part of the brain that converts short term memories into long term memories. So far, the chip has only been connected up to a rat brain. There's still a lot of work to be done before it's ready for use in humans, but it looks like they are making good progress. For the rats, at least, the chip does seem to operate like a real hippocampus.

"Flip the switch on, and the rats remember," says Berger. "Flip it off, and the rats forget."[8]

The next step will be to start testing with monkeys instead of rats, with the eventual goal of being able to help humans who are suffering with Alzheimer's disease or whose brains have been damaged due to stroke or injury. If it can be shown that these chips can really restore a person's ability to remember, there are still some other challenges that must be met before they can start being used as medical implants. Connecting electrical wires to neurons is not a simple matter involving an electrician's crimping tool, nor does the brain come with a set of terminal blocks. Instead, the usual method is to push thin wire electrodes down into the squishy cortical tissue, and hope that they make contact with the right neurons. The dendrites and the wires do not match up in a one-to-one fashion, like in a wiring diagram, since the electrodes are not nearly as thin as neurons. Instead, with a little less finesse than you'd find in a typical electronic circuit, each electrode is used to stimulate all of the neurons that happen to be in the vicinity.

Fortunately, says Berger, experience with cochlear implants shows these kinds of brain prostheses can still work even if the signaling is rather crude.[9] Thankfully the brain usually grows accustomed to the implant, and learns to make the most of the new sensory input. Cochlear implants are used to restore the ability to hear, and they work by directly stimulating the auditory nerve. These bionic ears have been installed in more than 200,000 people, making them the most widely used brain prostheses to date. But although they've been around for several decades, they are still far from perfect, and cannot, as yet, provide normal hearing ability.

As you'd imagine, the wet environment of the brain is not the best place for metal wires. Electrodes can corrode, or become dislodged, and the constant electrical stimulation can kill some of the adjacent neurons. The patient's cells may also mount an attack on the foreign intruders, which can lead to the buildup of scar tissue. All of this can cause a reduction in signal quality over time, depending upon the electrode configuration and the voltage levels that are used. Over the past decade or so, electrodes have gotten smaller, and they can be inserted with greater precision. Engineers are continuing to look at other possible improvements, such as using ceramics or composite materials to make the electrodes stronger and less likely to corrode.

On the east coast, there's a similar research effort under way, in a collaborative program called BrainGate, involving Brown University in Rhode Island, Massachusetts General Hospital, and the Providence VA Medical Center. The BrainGate technique also involves the use of electrode arrays, but in this case the electrodes are implanted in the motor cortex rather than the hippocampal region. It was announced in 2011 that a paralysis patient with the BrainGate implant was able to interact with a computer, using her thoughts to move the cursor around the screen. She had electrodes implanted in her head for almost three years, and although there was some signal degradation during this time, it gives researchers reason to be hopeful these kinds of prostheses may be viable as long term solutions.

Sometimes the electrodes themselves are all that's required to restore functionality. Like jumper cables for a car, they can provide the necessary electrical impulses to get the brain working again, or to suppress involuntary actions. It's called Deep Brain Stimulation (DBS), and requires the use of a pacemaker to regulate brain activity by delivering electrical signals to specific locations via the implanted electrodes. The approach has been used for a number of years to help quell the motor tremors caused by Parkinson's disease. In 2007, The Washington Post described how DBS helped a debilitated young man regain his ability to speak, after being minimally conscious for six years. The man was thirty years old when he was attacked and severely beaten on his way home, and left with serious head injuries. Doctors told his parents that he would probably remain in a vegetative state. However, with help from the DBS device, which now stimulates his brain for a twelve hour period each day, he has regained some of the functionality that was lost, including the ability to chew his food, move his arms, and respond to questions.[10]

Electrodes are not the answer to everything, of course, and in particular we'll need to find other ways to fend off the neural atrophy that is a characteristic of diseases such as Alzheimer's and Parkinson's. To that end, although brain prostheses have a lot of potential, it will be equally, if not more important to find solutions involving biochemistry, rather than electronics or exotic materials. This might include new medication that can regenerate neurons, or repair defective neural branching. Even when prosthetics are used, it may be in combination with regenerative drugs that can optimize their effectiveness. The coming decades will surely see tremendous progress on all of these fronts, but we may have to wait a lot longer before we see positronic brains starting to roll off the assembly line. There's still a lot more work to be done first with our old fashioned, organic noggins.

Building a conscious machine is probably the single greatest technical challenge that anyone has ever dreamt up. Nevertheless, according to Kurzweil, we'll have the job done by 2045. But what about that kick-ass brain scanner that we're going to need? Not to worry, says Kurzweil. The new scanners will be out in the 2020s. The solution, he says, will be to use nanotechnology to build tiny robots that can swim around inside our brains, and like miniature versions of Lewis and Clark, they'll be able to map out all of our neural pathways and connections.[11] One problem, though, is that these nanobots haven't been invented yet, and nanotech expert Richard Jones is not even sure that the concept is sound. He believes that such tiny mechanical contrivances would fall apart, or become otherwise incapacitated in the gooey, microscopic world of the cell.[12]

So far, the most talked about medical application of nanotechnology has been the so called lab-on-a-chip concept – microchips that can be used to detect bio-warfare agents, or to quickly identify disease markers in a drop of blood. Researchers are also looking at using nanoparticles to deliver drugs to their intended targets; a technique that could be used, for example, for more precise targeting of cancerous tumors. Nanoparticles are now being used in all kinds of commercial products and some scientists are concerned, since little is known about the carcinogenic properties of these new materials. Unfortunately, it's not unheard of for manufacturers to create products that cause more problems than they solve, so a bit of caution might well be warranted.

Kurzweil assures us that nanotechnology will be the answer to all of our problems, and that we'll be able to use nanobots to construct all of the material objects we desire, including new bodies for our artificial brains.[13] Searle is highly skeptical, and says of Kurzweil's work:

> I cannot recall reading a book in which there is such a huge gulf between the spectacular claims advanced and the weakness of the arguments given in their support.[14]

Journalist Glenn Zorpette points out that Kurzweil is not alone in making unfounded assertions about the future. Many singularitarians are particularly fixated on the idea of uploading brains into computers. They believe not only that this will be possible, but that it will happen, conveniently enough, within their own lifetimes.

> Most of all, we note the willingness of these people to predict fabulous technological advances in a period so conveniently short it offers themselves hope of life everlasting.[15]

In other words, this unrestrained optimism is just another example of humans grasping at straws in an attempt to find an escape from death. The singularitarian philosophy, it seems, is that a technological solution to death is just around the corner, if only we can hold on to life for just a bit longer. To that end, Kurzweil teamed up with physician Terry Grossman to write a guide book that explains how to "live long enough to live forever."

As you might have gathered, I'm not a big fan of death myself, and I don't have anything against the denial of death per se, but neither do I see any point in being unrealistic about it. Unfortunately though, the singularitarian creed appears to be little more than wishful thinking. Granted, nanobots may sound a little more plausible as a strategy for immortality than building a pyramid or getting monkey testicle implants. Nevertheless, I would rate Kurzweil's chances of being replicated in nanotech heaven as being pretty close to zero. I suspect he has his own doubts, which could explain his interest in Alcor; an outfit that allows you to have your body frozen when you die, in the hope of being resuscitated by some future technology.[16] Ungrounded faith in the potential of technology may provide a more hopeful alternative for those who have trouble believing in God, but it will be no more effective than religion in actually staving off death.

There's no question that technology will continue to change the world, and hopefully for the better, but the rate of progress may be slower than singularitarians are willing to admit. As far as brain science is concerned, Ted Berger's artificial hippocampus is about as cutting edge as it gets, and yet this device really just facilitates a fairly crude propagation of action potentials. It's nowhere near as complex as an actual hippocampus. How then can we account for all this optimism regarding artificial brains? A lot of it stems from the astounding growth of the computer industry over the past few decades. The miniaturization of electronic components has been the driving factor behind this growth. As engineers figured out how to squeeze more transistors into smaller spaces, our computers have grown ever more powerful without getting more expensive.

In 1965, engineer Gordon Moore predicted that the number of transistors in our processing elements would double every two years; a prediction that turned out to be pretty accurate in the decades that followed. Moore's Law, as it came to be known, is often used by singularitarians to justify their belief in exponential growth. Kurzweil mentions it in his book, although he also comes up with his own law, called the law of accelerating returns, which forecasts a continuous exponential growth in technological capability. Medical technologies will see "a doubling of capacity each year," according to Kurzweil and his coauthor Terry Grossman.[17] "The spatial resolution of brain scanning is doubling every year, and the amount of data we are gathering on the brain is also doubling every year."[18]

Yet, even Gordon Moore himself doesn't believe we're headed for a singularity.[19] Assuming that the number of neural connections in the brain bears some analogy in terms of processing power to the number of transistors in a computer, Kurzweil predicts that computers will be as powerful as brains by 2020.[20] But is there a magic number of transistors that takes us closer to being able to duplicate the causal powers of the brain to produce a conscious mind? Harvard psychologist Steven Pinker doesn't think so.

> Sheer processing power is not a pixie dust that magically solves all your problems.[21]

It is understandable that information technology has generated so much optimism and exuberance. When I was a programmer back in the eighties, we had the same creative potential as today's software developers, but we were severely limited by the available hardware. We had to be conservative, using as few bytes as possible to store each piece of data. Once computers got a little more muscle, developers grew extravagant and began using up memory and processing power like there was no tomorrow. All that pent-up creativity was released, resulting in the worldwide web of software that has had such a transformative effect on our lives in the past couple of decades. Moore's Law was not really a law, but more of a standard by which performance was measured and targets were set. However, unlike angels on the head of a pin, there is a limit to how closely you can squeeze things together on a chip, which means the chip makers will need to continue to look for innovative ways to build more capable machines.

Information technology aside, there are many who argue that technological growth hasn't been as dramatic as futurists have been predicting. As I remember the car that I drove back in the eighties, I can't say that it was any less fun to drive than the one I have now. Just before her eleventh birthday, my daughter told me that her most favorite movie of all time was *Back to the Future*, another product of the eighties. We watched that movie recently, and my kids didn't complain that the movie was dated, or that the special effects were lame. I'm not sure that I would have felt the same about a thirty year old movie when I was a kid.

As you may remember, in *Back to the Future II*, the protagonists travel forward in time, to the year 2015, where they find flying cars, hoverboards and home fusion devices. It's now 2012, and there's still no sign of the future. Sometimes it feels like we've left our most futuristic technology back in the past, as we reminisce about the old space shuttle fleet, for example, or supersonic airliners, or a more distant era when people traveled to the Moon. I overheard my daughter telling my five year old son that it was pointless wanting to be an astronaut when he grows up, because nobody wants to go into space anymore. It's a fact that we have flat panel displays, GPS navigation, tablet computers and smart phones, but the truth of the matter, says philosopher and science historian Alfred Nordmann, is that his grandmother saw more progress in her lifetime than he's seen in his.

> She witnessed the introduction of electric light and telephones, of automobiles and airplanes, the atomic bomb and nuclear power, vacuum electronics and semiconductor electronics, plastics and the computer, most vaccines and all antibiotics.[22]

Although we may not see flying cars, sentient machines, or Martian colonies in the near future, that doesn't mean we won't have a bright new tomorrow. There are still plenty of examples of science fiction becoming reality, such as the new artificial hands from Touch Bionics in Scotland. With their cool looking skin, and movable fingers, we're finally starting to see the sort of technology that was used on Steve Austin back in the seventies. I don't think I'd be going out on a limb to suggest that we'll continue to see bionics improve over the next couple of decades. They may not come with atomic power packs, and the warranty might not cover you if you decide to leap off a tall building, but there's no reason we can't have bionic arms and legs with the same look, feel and functioning as the regular flesh and bone variety.

It's also possible that, perhaps by the end of the century, we might be able to reconstruct a missing limb without the use of electronics or synthetic materials. When a salamander loses an appendage, it's actually not that big of a deal. It simply grows a new arm or leg to replace the one that's missing. In theory, the same approach could work for humans as well. We'd need a way to control the biochemical signaling that occurs at the injury site, so that instead of just covering the wound with scar tissue, the cells are given a green light to start rebuilding the missing body parts.[23]

The coming century should also see some significant gains in the war on cancer, as so many new battle plans are being developed. In many cases, scientists are working on techniques to prepare the body's own defensive systems for more effective combat against malignant tumors. For example, some drugs are designed to stimulate a response from the body's T-cells; immune system cells that hunt down foreign invaders and kill them, just like hired assassins. The problem is that tumors usually consist of cells from the patient's own body, so they are not easily recognized as foreign. By helping to unveil the villainous nature of these tumor cells, and by properly training the T-cells to recognize them, we make it easier for the body's hit squad to take them out.

An example of this approach hit the headlines in 2011, when NBC Nightly News announced some extraordinary results from a study at the University of Pennsylvania involving just three leukemia patients who were treated with an experimental new technique. Doctors removed some of the patients' own T-cells, and then used a virus to genetically reprogram them to attack cells that showed signs of being cancerous. They were also programmed to make more copies of themselves once they began encountering their prey. The modified T-cells were then put back into the patients where they began acting like a SWAT team in eliminating tumor cells. In two of the patients the cancer was completely eliminated, and in the third it had shrunk by seventy percent. Hopefully we're beginning to see the writing on the wall, and maybe by the end of the century we'll have successfully dealt with most, if not all cancers.[24][25]

I enjoyed writing science fiction stories as a kid, and it was something that the grownups encouraged as there was nothing wrong with a little imagination and curiosity. These were highly valued traits that could lead to greatness in society. We were taught to admire the great icons of change, such as Alexander Graham Bell and Thomas Edison, and we learned to ridicule the ignorance and stupidity of those who tried to stand in the way of progress. For me, that was the point of science fiction. It was a celebration of technology and a ratification of our ideology of change. Things are not supposed to stay the same. Parents expect their kids to have opportunities that they never had. Some see a hopeful future, while others see dystopia, but we all see change. Sometimes the changes have a profound effect on human history, such as when we decide to link the coasts with a transcontinental railroad, or send space travelers to the Moon. Non-stick pans and wrinkle-free shirts might not have the same gravitas, but they provide a welcome relief from tedium. New gadgetry is not only useful, but fun to play with.

There are varying opinions as to why change is a good thing. Building the railroads made good business sense, and helped bring people together. Space exploration is clearly something we are destined to do, just as we were compelled to explore the west. Exploration must be a good thing, because without it the United States wouldn't exist, and I wouldn't be living in California. Even as we complain that our products should be more recyclable, and should use fewer toxins, we know that we can't live without them, and we know that if we don't keep making products, we'll all become unemployed.

For better or for worse, we are married to our technology. Whether trivial or profound, it has changed the nature of our civilization and our relationship with the rest of nature. The change that we bring to the world is usually well intentioned, but there have been some negative consequences, such as air and water pollution, and adverse effects on public health; the loss of biodiversity and destruction of natural habitats; and the inadvertent disruption of the global climate. There is a mindset that speaks for the noble purpose of humanity, arguing that despite these negative influences, our technology, on balance, has made the world a better place. My dentist, for example, is pretty happy with global warming, and says that it has to be better than global freezing. Even for those who hold a different opinion, the usual consensus is that, for the most part, the problems we have caused with technology can only be solved by technology.

When I consider the ways in which technology has made a difference, the solutions that matter most are those that allow people to live, when they otherwise wouldn't. In the 1920s, America was losing fifteen thousand people a year to diphtheria, which is now no longer an issue.[26] Appendicitis once meant an agonizing death, whereas today it is little more than a couple of weeks of down time. Our philosophy of progress through science needs no further justification other than the fact that it can save lives. Until someone comes up with a better strategy for accomplishing miracles, the matter is settled. We all know that life is short, but sadly for some people, it is even shorter than it should be. The most intolerable reality is that people are still dying before their time. I had a friend who died in his twenties, another in his thirties, a work colleague in his forties, and an uncle who was struck down at age fifty eight. I would gladly live without flying cars and hoverboards if it meant that by the end of the century we could stop losing people to cancer. In my book, that's what our technology needs to be about – keeping our family and friends around for as long as we can.

7

⌘

God's Country

The kids' library at Dolphin's Barn was where I first found out that the Sun was going to die. As a kid, I found this a little disturbing, since I knew that if the Sun died it would also be bad news for Earth. Thankfully, the book that I was reading explained that this wouldn't happen for another five billion years, so there was no need to panic. I could tell that five billion years was a long time, and I was certain that by then humans would have colonized the distant reaches of the galaxy. So when it came time for the Sun to kick the bucket, nobody would be harmed because everyone would have moved on. In any case, Granddad told me not to worry about the future inhabitants of Earth, since he and I would be dead by then.

Even though I knew that I'd be long forgotten when it happened, it still made me a little sad to think of our planet's final days. However uncomfortable we feel about our own personal mortality, I guess it would be worse to know that civilization itself was doomed. If the human race can be said to have any long term goals, survival must necessarily be at the top of the list. That was how Charles Darwin felt about things as well. Darwin believed that we were destined for greatness, and it bothered him to think that everyone would be wiped out of existence in due course.

> Even personal annihilation sinks in my mind into insignificance compared with the idea, or rather I presume certainty, of the Sun some day cooling & we all freezing. To think of the progress of millions of years, with every continent swarming with good & enlightened men all ending in this.[1]

> Believing as I do that man in the distant future will be a far more perfect creature than he now is, it is an intolerable thought that he and all other sentient beings are doomed to complete annihilation after such long-continued slow progress.[2]

At that time, you see, British scientist William Thomson, who would later be known as Lord Kelvin, knew that it was only a matter of time before the Sun ran out of fuel, at which point Earth would be plunged into an icy darkness and all would perish. However, not being well versed in nuclear physics, he greatly underestimated the amount of time we had left and so he told everyone that humanity would expire in just a few million years.[3]

Even though he got the date wrong, Thomson was essentially correct in pointing out that our solar system is destined to become a cold and lifeless place, and if we look far enough into the future, we see that the same eventual fate awaits the entire universe. In 1929, astronomer Edwin Hubble was able to convince the world that the universe is expanding, which means that the farthest galaxies are getting even more distant. That fact was surprising enough, but then in 1998, two separate research teams discovered that this expansion was not slowing down, as everyone expected. Instead, it was actually accelerating. As if the rest of the universe wasn't far enough away already, this stretching of space means that all the galaxies have put their gas pedals to the floor. They will continue to accelerate away until they are out of view and completely undetectable.

The long term prospects for life are thus quite bleak. From the point of view of our descendants 100 billion years from now, there'll be nothing left in the universe apart from our local group of galaxies, which by then will have merged into a supercluster. We will be entirely alone in an endless ocean of blackness. In about 100 trillion years our local stars will have exhausted their fuel supplies, and the supercluster will then eventually collapse to form a black hole. At that point the show is over.[4]

Not so fast, says Ray Kurzweil. Long before the universe fizzles out, our super intelligent machines will have taken over the universe, and they'll be able to engineer a better outcome. In the distant future we'll be so powerful that we will no longer be limited by mere laws of nature, which means, says Kurzweil, that the fate of the universe is not cast in stone, but rather a decision to be made by our intelligent offspring.[5] So, that sounds a bit more promising.

Physicist Frank Tipler also predicts that our descendants will become godlike in the distant future. He suggests that the best outcome for the universe would be for it to stop expanding and then collapse, so that all the galaxies come together again. In this scenario, he expects that our distant descendants will harness all of the matter in the universe and use their awesomeness to become truly immortal. We'll all then be emulated in some advanced computer system, says Tipler. Our computational power will be so immense that it will even be possible to resurrect the dead, by emulating all the human beings that could ever possibly have existed.[6]

Unfortunately though, the universe doesn't show any signs of wanting to collapse, so it doesn't look like Tipler's plan is going to work. Suffice it to say that if the universe is ultimately doomed, then immortality is out of the question. Even Tipler admits that an eternally expanding universe means curtains for life.

> It would be all over for life. It would be all over for any possibility of having purpose in the universe.[7]

So unless some future deities can work a great deal of magic, we can forget about living forever, or resurrecting the dead. I hate to say it, but it looks like Granddad Waine is gone for good. I'm also starting to think that we're not going to get anywhere by speculating about what might happen in 100 billion years. As I recall, Gerry Edelman said that anyone who tries to go beyond ten years is a crackpot, so I'm worried we might have gone a little too far over the threshold. Besides, when people ask for immortality, are they really expecting more than 100 billion years? *Star Trek*'s Lieutenant Uhura was offered immortality in the form of an android body, but even when she learned that the new body was only good for 500,000 years, she still seemed happy enough. 500,000 years is not bad, when you consider that most of us have to make do with less than a century.

I remember giving an astronomy lecture in which I described how planetary rings, such as those around Saturn, might be a relatively short lived phenomenon – perhaps lasting no more than about 100 million years.[i] For planetary scientists, that's hardly any time at all. To a planet such as Earth, 100 million years feels about the same as a year does to us. Maybe if we had the lifespan of a planet we should probably start singing *Auld Lang Syne* at the end of every 100 million year period, instead of every year.

It's incredible, really, how little time we have. In April 2011, as President Obama was getting ready to meet Japanese leader Yoshihiko Noda, commentators noted that in the previous four years Japan had no less than four prime ministers. This lack of stability in Japan's political system was apparently causing some headaches for the Obama administration, by making it very difficult for them to have a continuous dialog with their Japanese counterparts. I'd imagine the same sort of problems would arise if we were to engage in communication with a nearby alien civilization, say 100 light years away. Our short lifespan would be very frustrating for the poor aliens, making it almost impossible for them to have any kind of a meaningful conversation with us. Before they'd get a chance to respond to a question, the people on the other end of the line would all be dead.

So the next time someone wants to know if it's possible to live forever, maybe you should suggest that they lower the bar just a little. Bumping up the lifespan by a few hundred years might be a reasonable goal, but asking for immortality definitely qualifies as overreaching. I've recently seen two separate TV shows entitled "Can We Live Forever?" which, I'll admit, sounds more dramatic than "Is there a way to stave off death indefinitely?" However, the latter question is probably more appropriate.

Unfortunately, as author Samuel McCracken reminds us, the progress of civilization has done nothing to increase the human lifespan.

> It seems clear that the maximum useful lifespan has not varied significantly in historical times. The evidence is clear from the case of Sophocles and his sons that in 5th century Athens 92 was thought a great age and yet it was an age at which an exceptional man might function exceptionally.[8]

---

[i] New evidence from NASA's Cassini probe suggests that Saturn may have had its rings since the early days of the solar system.

In other words, there have always been people who managed to live to the limits of old age. McCracken provides the example of Sophocles, a playwright from ancient Greece who, at the ripe old age of ninety-two, was brought to court by his sons who argued that the old man was not of sound mind. The boys were fed up waiting for him to die and wanted what was coming to them. In his defense, however, Sophocles presented the court with the play he had just written – the *Oedipus Coloneus* – whereupon they ruled in his favor, finding him to be in full possession of his faculties.

In the long term, as we've seen, our planet's days are numbered, but what about in the short term? Is there any danger of Earth getting destroyed in the next few hundred years or so? That would certainly put the kibosh on any plans to extend human life. The good news in that respect is that our home world has been around for a very long time indeed. Life got a toehold soon after Earth was born, and it has persisted without significant abatement ever since. The past half billion years in particular has seen a bewildering explosion in the diversity of living things. Biology has overwhelmed the planet; from the deep oceanic vents to the dry deserts of Antarctica, every conceivable niche has been occupied. There have been a few cold spells, volcanic eruptions and such, but otherwise life has gone about its business, relatively unperturbed.

That being said, if the dinosaurs could have left us a written record of their experiences, I am sure it would attest to the fact that getting hit by a ten kilometer wide asteroid is no fun. Imagine a giant mountain of rock hurtling toward us at twenty kilometers per second. The resulting explosion would create a huge fireball, and would throw so much debris into the atmosphere that we'd all be living with the consequences for years afterwards. That's what we believe the dinosaurs had to contend with, some 65 million years ago. What if an object of similar size is headed our way as we speak?

Luckily, impacts like that don't happen very often. Most of the space rocks that hit the atmosphere are of the smaller variety and usually go unnoticed, but an event similar to the one that upset the dinosaurs would impose a devastating cost on humanity. Most of us would probably escape the initial blast, given the extent to which our population is distributed around the globe. However, we would then need to deal with the fiery ejecta from the massive explosion, which could set much of the world ablaze. The shock from the impact could also trigger earthquakes, volcanos and tsunamis. Beyond these immediate dangers, we'd then be faced with several months of darkened skies due to all of the dust kicked up into the atmosphere. A thick aerosol blanket would severely limit the amount of sunlight reaching the surface, which would hamper photosynthesis and thus play havoc with our agriculture. It could take years for the air to clear. Millions would go hungry, and widespread suffering would ensue.

As terrible as all this sounds, it is not, however, a foregone conclusion that we would go the way of the dinosaurs. We are a resourceful and adaptive species, with a demonstrated ability to survive in all kinds of environments, and a versatility that allows us to meet new challenges. It is more than possible that we might pick up the pieces and rebuild our civilization. But the biggest advantage we have over the dinosaurs is our ability to predict these kinds of events, and to take measures to prevent them from happening in the first place. If there is a big asteroid heading in our direction it's quite possible that we might notice it, and with enough lead time we might even be able to give it a little nudge, so that it misses us completely. This kind of catastrophe is entirely preventable, as long as we keep a diligent eye on the heavens.

Global warming, on the other hand, is no longer preventable, so that's something we're going to have to struggle with. It's difficult to predict exactly what will happen in the long term, although scientists can estimate the most likely outcomes. It's generally expected, for example, that sea levels will continue to rise, and that a significant increase is definitely on the cards at some point. All we need is for some big blocks of land ice to slide into the ocean, and that should do the trick. At this point, most of our ice resides in Greenland and Antarctica, where it builds up over time to form thick sheets that then creep gradually toward the sea. These ice flows are slow moving, partly because they are buttressed by walls of sea ice. But as the sea ice disintegrates, there is a concern that the flow of ice into the ocean will speed up, which could cause a significant rise in sea level. For the sea to rise, the ice doesn't even have to melt. No one yet knows how quickly this will occur, but sooner or later we can expect to see increased flooding in our coastal cities.[9]

Another fairly certain prediction is that as temperatures continue to rise, tropical diseases will become more prevalent at higher latitudes. Species that are unable to adapt to the changing climate will die off, and ecosystems will collapse. Storm systems will likely become increasingly violent due to all the extra energy in the atmosphere. Desert regions are expected to expand, and water shortages will become more pronounced. To make matters worse, melting permafrost will release huge quantities of methane, which will exacerbate the atmospheric warming. Scientists cannot precisely say how this warming will affect climate patterns and ocean currents, but the effects will make themselves known in due course. As these disruptive changes are still unfolding, we may need to wait a century or two before the true cost of global warming is understood in terms of the toll it exacts on human lives and the global economy. If there's a plus side to all of this, we can at least suggest that while the going will get tough, it's unlikely that global warming will destroy the planet, or wipe out humanity. That's assuming, of course, that we eventually get our greenhouse gases under control, so that we don't end up creating another Venus.

Other than a head on collision with an asteroid, there are no other obvious dangers from outer space that we need to worry about. There are no Martians, which rules out the scenario from *The War of the Worlds*. To suppose a threat from alien invaders, one would first have to assume that aliens exist, which is a big assumption, and that they have the wherewithal and the inclination to traverse the vast distances between the stars. Our planet is situated in the outskirts of the galaxy – in the boondocks, as it were – so the aliens are unlikely to stumble upon us by accident. Unless warp drive is possible, which is another big "if," it could take them hundreds of thousands of years to get here, assuming they had some reason to select our star system from the list of several hundred billion possible destinations.

Since we are not yet capable of interstellar travel, the aliens would necessarily have superior technology, and since we only have about five thousand years of recorded history, their civilization would in all likelihood be much older and wiser than ours. It seems somewhat ludicrous then, to assume that such a civilization would go out of their way to destroy humanity. What possible motivation might they have? Other than the uniqueness of our biology, there are no raw materials on Earth that couldn't be found elsewhere in abundance. We have no products or resources that advanced aliens wouldn't be able to fabricate for themselves. With all things considered, I think it's a safe bet that aliens will never show up, benevolent or otherwise, although it would be interesting to be proven wrong.

A supernova explosion would pose a very real threat, provided it was close enough for us to feel the effects of the radiation it produced. Fortunately for us, nearby events are highly improbable in the near future. But if any of our neighboring stars are determined to go down this road, it's not like there's anything we can do about it, so it's probably best not to dwell on this any further.

What about the possibility that we might destroy ourselves? We still have more than enough nuclear weapons to get the job done, if we all put our minds to it. Former Secretary of State William Perry tells us that for him, the prospect of nuclear war was always terrifyingly real.

> I'd get calls in the middle of the night telling me that the computers are showing two hundred missiles on their way from the Soviet Union to the United States. This really gets your attention.[10]

Figure 8. Hiroshima, Japan, after an atomic explosion. [11]

Perry got together in 2012 with former Secretary of State George

Shultz and former US Senator Sam Nunn for a meeting of The Commonwealth Club of California in San Francisco, where they discussed the current state of the nuclear threat. While they all seem to agree that the chances of a global nuclear exchange are quite remote, there's still a very real possibility that a terrorist organization could get hold of a nuclear device and use it to obliterate one of our cities. Apart from the terrorist threat, Perry also worries about the nuclear standoff between India and Pakistan.

> That's a scenario of how we could not just have one nuclear bomb in a city, a terror bomb, but we could have a nuclear war break out between two major countries of the world.[12]

"Perhaps the probability of the use of these weapons is low," admits Gloria Duffy, President of The Commonwealth Club of California, "but the consequences of that low probability event are unimaginably horrible."

Fortunately our political leaders now seem to understand the danger, and they are working diligently to prevent the spread of nuclear materials. Nunn tells us that "The most important thing is keeping that nuclear material out of the hands of terrorists."[13]

That's why we've been buying enriched uranium from former Soviet countries and using it to generate electricity here in the US.

> Twenty percent of our electricity in America comes from nuclear power. Fifty percent of that material that we burn today in America's nuclear power plants comes from highly enriched uranium that was in the form of bombs aimed at us during the Cold War. So by definition, ten percent of America's electricity represent in the biblical phrase, swords that have been turned into plowshares.[14]

The ultimate goal is to rid the world of nuclear weapons, says Nunn. As naïve as this may sound, he believes that it is an attainable goal, although it will take many years and will require lots of small steps. After decades of cold war tension we can finally start to breathe again, and hopefully put those apocalyptic visions of the future back into the past, where they belong.

But hold on there just a minute, I hear you say. Isn't it only a matter of time before some engineered virus breaks loose from a laboratory and goes on the rampage? Hopefully nobody would deliberately create something with the ability to spread throughout the biosphere and eradicate all human life, but is that something that's even possible?

In essence, a virus is a just piece of genetic code (DNA or RNA) that's covered over with a protein coat. As such, viruses don't really live by themselves since they depend upon the biological resources of organisms in order to replicate and survive. You can think of a virus as software, in which case organisms are computers that can execute the software. Without a computer, the software code is harmless.

In the normal course of events, our cells diligently carry out the instructions from our own DNA library. If a virus manages to dock with one of these cells, it will inject its genetic code through the cell membrane into the interior of the cell, where it gets picked up by the cellular machinery which then dutifully carries out the viral instructions. In this way, the cell can be coerced to make more copies of the virus, which then exit the cell to spread their mischief elsewhere.

Viruses then, are all about bits of genetic code getting passed around and exchanged between organisms. The code changes and evolves over time, due to the cumulative effect of mutations. There are, of course, other ways in which genetic code gets shared and transformed, with sex being perhaps the most popular. Bacteria are particularly fond of swapping and exchanging genes, which is how they can evolve so quickly compared to the rest of us. In truth, this genetic interchange process is really what biology is all about.

This means that when we talk about genetically modified organisms, or engineered viruses, we're not really talking about something new. Nature has been at this for quite a while already. Earth's biosphere is, in effect, a giant natural laboratory, constantly churning out new genetic variations, and interfering with the software of life. Whenever we see the emergence of a new virulent infection, it's because nature can't stop meddling with the code.

From our perspective, the most frightening outcome would be a virus that is both lethal and contagious, but these two traits don't always work well together. If the virus is too effective as a killing machine, the host may die before the infection has a chance to spread. For this reason, as evolutionary biologist Paul Ewald explains, the most dangerous pathogen is one that is either carried by a vector, such as a mosquito, or one that can survive outside the body until it gets a chance to infect a new victim.

> Infectious agents that survive for weeks or years in the environment tend to cause death most frequently: for example, smallpox kills one in 10 people, and the virus that causes it can persist for more than a decade outside a host.[15]

As much as we might worry about what the next mad scientist will come up with, the fact is that nature is the real expert at designing new pathogens. Even if an engineered agent is proven effective in a lab, it still needs to be able to compete for survival in nature's arena.

It's also worth mentioning that the whole mad scientist theme may be a bit overplayed. The truth is that scientists are often the most cautious in dealing with discoveries that could potentially put people in danger. In 1975, for example, scientists took the initiative and organized a conference at the Asilomar center in California to talk about the possible consequences of genetic engineering. And even though there was no obvious cause for concern, they agreed to self-imposed restrictions on certain kinds of research until the risks were better understood. That's arguably a better standard of behavior than you'd get from a bunch of industrialists or politicians.[16]

The suggestion then is that engineering an infectious agent to exterminate humanity might not be as easy as one would expect. Be that as it may, I don't want to suggest that there are no evil scientists or that it's impossible to manufacture a deadly pathogen; nor would I like to see my hypothesis being put to the test.

There's little doubt that the potential for nuclear war was the scariest threat we've suffered through so far, at least with respect to global catastrophes. Now that this threat has diminished, there's room for optimism regarding our prospects. For some, though, there is still a general sense of unease about the future, a feeling that perhaps someday we might stumble accidentally upon some new technology that will blow us all to smithereens. Maybe that's why we're not seeing any alien civilizations, because once they reach a certain level of technological sophistication, they invariably self-destruct.

Perhaps a science experiment will go awry, setting off a chain reaction that will destroy the biosphere. That's the sort of thing that people worried about when scientists told us what they were planning to do at the Large Hadron Collider (LHC) in Europe. The plan, they said, was to use this massive particle accelerator to recreate the energy levels that existed at the moment our universe was created in the Big Bang, by smashing particles together at extreme velocities. Nobody had ever done this before, and that's the part that got people worried.

Scientists said there was nothing to be concerned about because the colliding particles were extremely small, so the LHC would only create miniature Big Bangs as opposed to the kind that could destroy the universe. So far it looks like they were right, as the experiments don't seem to have caused any serious problems.

As trustworthy as our scientists may be, this is an example of why it is important for the rest of us to know something about science, so that we can make some of these assessments for ourselves. Otherwise, says astronomer and science evangelist Carl Sagan, we're sure to have lots to worry about.

> We've arranged a global civilization in which most crucial elements – transportation, communications, and all other industries; agriculture, medicine, education, entertainment, protecting the environment; and even the key democratic institution of voting – profoundly depend on science and technology. We have also arranged things so that almost no one understands science and technology. This is a prescription for disaster.[17]

The future is clearly not without risk and there are challenges to be faced, yet no one can definitively say that we are destined for doom and destruction, or that civilization is in its twilight years. Our destiny is still very much up in the air. Fear mongering is part of our nature, but it seems to me that for every apprehension, we have individuals working hard to mitigate the risks and improve our situation. We have hardships to manage, but with the right frame of mind they appear no worse than ones that we coped with in the past. I tend to have an optimistic outlook, perhaps due to the inspiration that I get from watching my kids grow up. I am impressed at how capable they are becoming, and when I get to see the world through their eyes, the future looks brighter than ever.

One noticeable point, however, is that science fiction often portrays us in a rather dim light. A recurring theme is that humans are a backward and barbaric race, and that we set a bad example for the rest of the galaxy. The aliens give us a severe talking-to for being so awful, and to justify their disapprobation they trot out the usual stock footage from World War II, showing cities being destroyed, missiles being launched, and people being ill-treated by Nazis. The humans never seem inclined to challenge this point of view. Instead, we take it as a given that humans behave disgracefully, but try to make the point that we have a lot of other things going for us. Even Granddad Waine would often have harsh words to say about humanity, as he talked about "man's inhumanity to man," and how we were the only species that preyed upon itself. But are we really that bad?

When we consider the atrocities committed in the past, we must acknowledge the sad reality that humans are capable of committing horrible misdeeds. The Holocaust will remain a powerful and tragic reminder of the tremendous suffering that hatred and intolerance can cause, and of the ways in which religious, cultural, or political differences may be used to justify the most despicable treatment of others. Of course the Nazis weren't the only ones at fault. Every time I see Andrew Jackson's face on a twenty dollar bill, it brings to mind this country's legacy of shame, and the terrible plight of Native Americans affected by his Indian removal policies. Historian Howard Zinn reminds us of the four thousand people who perished during the forced relocation of the Cherokees, and how President Van Buren later informed Congress that things couldn't have worked out better.

> It affords sincere pleasure to apprise the Congress of the entire removal of the Cherokee Nation of Indians to their new homes west of the Mississippi. The measures authorized by Congress at its last session have had the happiest effects.[18]

As we remember all those who needlessly suffered throughout the course of history, it renews our resolve to eradicate such disgraceful behavior from society. No matter how advanced we believe ourselves to be, we must keep a watchful eye to prevent such things from ever happening again.

Nevertheless it would be wrong, I think, to characterize humans as barbaric and uncivilized. The truth is that if you were dropped at random into the middle of any major metropolis, instead of finding mayhem and carnage you're more likely to find yourself looking at a multitude of people going about their business in a peaceful and civilized manner. People are exceptionally sociable, and are uniquely skilled at living and working together in large groups. Frans de Waal of the Yerkes National Primate Research Center tells us that this isn't true of other species, and when it comes to playing nicely together, humans are much more civil than our nearest relatives, the chimpanzees.

> If you would put ten million chimps in New York City, you know, I don't think you would have the same kind of city. There would be … there would be a blood bath, basically.[19]

Far from being backward and barbaric, humans are the very definition of civilization. That's why our human brains evolved – to give us an edge with regard to social organization. We are, without question, the most compassionate, empathic and cooperative species on the planet.

Occasionally we see how much disruption can be caused by a single individual, such as the arsonist in southern California who was responsible for starting fifty four fires over a period of just four days in 2012. Fortunately these occurrences are so rare that they easily make the headlines, and it serves to underscore the fact that we are predominantly a peaceful race. We sometimes get frustrated when it takes us a long time to solve our social, economic and environmental problems, but with a global population approaching eight billion it's actually quite remarkable that humans are able to work together with any degree of coherence.

While we celebrate the human enterprise, we must, however, concede that things do not look so bright for the other animals with which we share the planet. As my African friends explain, it will no longer be possible to allow animals to roam about freely in the wild. It's too dangerous for a start, plus it gets in the way of progress. Global warming makes life even more difficult, as it allows some animals to move into new territories, thus putting pressure on the species that are already there. Other disruptive effects include water shortages and shifting migration patterns due to climate change, not to mention habitat destruction and environmental toxins introduced by human industry. Biologists warn that we are in the midst of a mass extinction, and that by 2100, fifty percent of our biodiversity will be gone.[20] Some species may be preserved in our zoos, or as biological samples in refrigerators, but many will be lost forever.

To summarize then, I think it is safe to say that humanity has a future, but it will not be business as usual. By the close of the century our supply of oil and coal will be almost gone, so we'll need to have made some serious progress in developing energy alternatives. Perhaps we'll have finally upgraded our electricity infrastructure and maybe even invested in a decent sized national solar array.[21] Over the next few decades we'll also start running out of fresh water, so that's another problem we'll have to face, particularly here in the western part of the country.[22] In the worst case, the cost of having to deal with these challenges may stifle progress in other areas, making it difficult to us to continue seeing the standard of living improvements we've grown to expect. That's not to say that we won't see an increase in economic growth, but this could be offset by the cost of having to support a larger population.

The upshot is that there's a penalty associated with our continued survival. Some might argue that life on Earth would be in better shape if humans weren't around, and that nature has been forced to pay too high a price for our existence. Others point out that we provide the means for the rest of nature to have meaning and purpose. Without us, the world would be pointless. But on the other hand, what is the point of humanity? As Sagan observes:

> There is no generally agreed upon long-term vision of the goal of our species – other than, perhaps, simple survival."[23]

As we reflect on the importance of our species, it's worth considering that, while the galaxy may have many life bearing planets, sentient civilizations such as ours may be very rare, or even nonexistent. If it's truly the case that we are the only species able to understand and reflect upon its own origins, then we owe it to the universe to survive and to reach our full potential, so that, in Sagan's words, "the cosmos [can] know itself."[24]

Setting aside the survival of the species, most of us, I think, would agree that nothing is more important than the safety and wellbeing of our families and friends.

"Taking care of yourself and your family – that's what it's all about," according to my dad.

I presume he meant as opposed to the hokey-pokey.

Of course, now that I am a dad, I've discovered that taking care of your kids is a complicated business. First you need to have all the necessary accouterments, such as BPA-free baby bottles, phthalate-free teething rings, PBDE-free furniture, organic mattresses, five-point harness car seats, bicycle helmets, a HEPA filtered vacuum cleaner, and locally grown, organic food. Your house must be free of asbestos and lead paint, and a monitored security system might not be a bad idea. Electrical outlets need to be blocked, sharp corners covered with foam, harmful substances placed out of reach, drawers and cabinets secured with safety locks, smoke and CO alarms installed, and dangerous areas fenced off.

Now before you say anything, I will admit that my wife and I probably reside on the more cautious end of the parenting spectrum. An Israeli colleague once told me that "You don't need to raise children. All you have to do is feed them, and they raise themselves."

He then proceeded to tell me how many times his two year old had broken his arm. Back in the eighties I had a manager who told me that there was no need to worry about hazardous substances. "The body has natural mechanisms for getting rid of toxins," he said.

That may be true, and perhaps we do overreact a little, but then again we would feel bad if our kids were found to be susceptible to some harmful chemical, or if they got injured because we weren't using the right kind of car seat. Perhaps we would relax if we had some assurances that manufacturers were careful not to use toxic materials in their products. Unfortunately, in the US at least, there are no such assurances. Government agencies are powerless to enforce safety, since most of the time companies are allowed to be secretive about the novel constituents of their products. A *Scientific American* editorial reported the following in 2010:

> Consequently, of the more than 80,000 chemicals in use in the U.S., only five have been either restricted or banned. Not 5 percent, *five*. The EPA has been able to force health and safety testing for only around 200.[25]

Senator John Kerry and his wife Teresa Heinz Kerry describe how the Europeans place the burden of responsibility on manufacturers to prove that their products are safe before they are released into the marketplace. In this country, however, new chemicals are assumed to be safe until people start dying, at which point it still requires significant effort on behalf of consumers to prove that the chemical was at fault.[26] As noted in the same editorial:

> Even when evidence of harm is clear, the law sets legal hurdles that can make action impossible. For instance, federal courts have overturned all the EPA's attempts to restrict asbestos manufacture, despite demonstrable human health hazard.[27]

In seventh grade, we were told that everything in nature was made of chemicals, which meant that all chemicals were natural. I think there was flaw somewhere in that reasoning, but it convinced us kids that there was no reason to be concerned about all the strange ingredients in the foods we ate. Partially hydrogenated oils, for example, were presumed to be perfectly safe.

Unfortunately, however, we presumed wrong. It turns out that the partial hydrogenation of vegetable oils produces trans-fats that are very bad for us. Not only does this stuff cause an increase in bad cholesterol (LDL) and triglycerides (fat molecules), it also lowers the good cholesterol (HDL), which means it significantly raises the risk of coronary heart disease – the biggest killer in America.[28] Even as this became common knowledge, the FDA insisted that partially hydrogenated oils were generally safe.

Effective regulation seems to make a difference. For example, a 2006 study in the New England Journal of Medicine found that an order of chicken nuggets and fries from a McDonald's restaurant in New York contained over thirty times as much trans-fat as the same order from a McDonald's in Denmark. That's because the Danes got rid of their partially hydrogenated vegetable oils by enacting legislation to severely limit the amount of industrially produced trans-fats in their food.[29][30][31] An improved regulatory system in the United States would not only save lives, it might also benefit American industry by bolstering consumer confidence.

The sad fact is that some of our engineered chemicals can kill people. Many diseases have now been linked to environmental toxins, so it stands to reason that we could save a few lives by adopting a more precautionary approach with respect to the environment. We are biochemical beings, so we are not immune from the effects of chemicals. If a synthesized agent has a molecular shape that is similar to that of a natural substance, then it could be worth investigating as a potential biohazard.

Things might not be so bad if we were careful not to allow our toxins to spill into the natural environment, but that seems to be easier said than done. In 2001, for example, it was announced that researchers working for 3M Corporation had discovered a chemical called perfluoro-octanyl sulfonate (PFOS) in blood samples taken from wild eagles in remote parts of the country. At the time, PFOS was used in Scotchgard, a stain repellent manufactured by 3M. Although the product was reformulated, PFOS is unfortunately a very persistent chemical. Ten years later, scientists at West Virginia University discovered that it was associated with chronic kidney disease in humans, and that it was now detectable in the blood of almost every American.[32][33] As my grandma used to say, "It's much easier to make a mess than to clean it up."

You'd wonder, though, why anyone would cause this much trouble over a stain repellent. Were we so concerned about staining our pants that we needed to add yet another pollutant to the environment?

In the past, nobody gave much credence to the idea that humans could actually influence the course of nature. America was a vast untamed wilderness, with unimaginable natural wealth. As historian H. Wayne Morgan recalls:

> Few believed that the earth's resources were intrinsically limited. . . . America rested on the idea of abundance. The possibility of obtaining affluence underlay the general adherence to individualism, and the aversion to governmental planning. To concede or accept the idea of ultimate scarcity would require a reevaluation of the national ethic.[34]

It was also believed that natural resources existed solely to be exploited by human beings. Nature served no purpose other than to further the immediate human agenda.

That kind of thinking might have been acceptable in the days when our impact on the environment was fairly minimal. However, in the period known as the Progressive Era, while America consolidated its position as the world's industrial powerhouse, some farsighted folks expressed concern that the unrelenting march of progress would eventually result in the reckless destruction of our natural heritage; although as Morgan points out, "The sheer scope of America inhibited any sense of crisis."[35]

Luckily for us, a few nineteenth century tree-huggers successfully prevailed upon Congress to start carving out pieces of pristine wilderness, making them off-limits to developers. Outside of these national parks, however, the subjugation of nature continued.

Now that the population has grown and the footprint of civilization has increased, we know that our traditional cavalier attitude toward the natural world is untenable. The Kerrys tell us that it's beginning to undermine our own safety.

> We have harnessed natural resources in positive ways – not just to feed, clothe, and house rapidly growing populations, but to increase the length and quality of life for hundreds of millions of people. But there's also no question that the careless exploitation of these same resources has begun to threaten the very foundations of life.[36]

We're also starting to realize the value of the environmental services that nature provides. In urban areas, for example, trees can lower the cost of air conditioning by providing cool shelter, they can clean the air, and they can absorb rainwater, which reduces the amount of money we have to spend on elaborate storm drains. By treating nature as an ally rather than a foe, we can improve the way we plan our communities by learning how to integrate more effectively with the environment.

Overall, society seems to be moving toward a greater appreciation of our connectedness with nature. There's a much greater emphasis on energy saving devices, and on reusing and recycling old products, as well as purchasing products that are made from recycled or recyclable materials. We're also getting better at handling hazardous waste, and businesses are finding that by making their products more environmentally friendly they can reduce costs while also generating sustainable growth.

It's probably fair to say that this new found respect for nature has been due, in large part, to the efforts of those people whom we like to call environmentalists. Scottish immigrant John Muir is often credited with the distinction of having been the first such creature in America. He used to live in Martinez, California, which is quite close to here, so I've been over to his house a few times. The first thing that you learn about Muir was that he was immensely passionate about the natural world. He had no time for the indoors, and spent a good part of his life roaming about by himself throughout the vast natural landscapes of the western United States.

Muir had a fairly intense religious upbringing. His father was a strict Calvinist, who believed in the importance of hard work and discipline. Like most people at the time, Muir would have seen the world through the lens of his religion. But unlike most people, he had no time for sitting in church and reading from the scriptures. Instead he had a strong spiritual connection with nature, and felt that this brought him closer to understanding the mind of God. Muir was notably awestruck by the natural vistas of Yosemite and the Sierras; a part of the world that he believed had more grandeur than any of the world's great cathedrals.

> Yet this glorious valley might well be called a church, for every lover of the great Creator who comes within the broad overwhelming influences of the place fails not to worship as he never did before. The glory of the Lord is upon all his works; it is written plainly upon all the fields of every clime,

and upon every sky, but here in this place of surpassing glory the Lord has written in capitals.³⁷

He became a respected naturalist, and also proved himself to be a talented writer. Stories of his adventures captivated the imagination of America, and were a driving factor in the decision by Congress to preserve Yosemite as a national park for the benefit of future generations. In 1903, President Roosevelt wrote to ask if Muir would accompany him on a trip to Yosemite, and so the two spent a few days camping out under the stars together.

Figure 9. John Muir and President Roosevelt on Glacier Point in Yosemite National Park.³⁸

Usually when I think about Muir, I imagine a man with a rather scraggly beard perched on some mountain shelf overlooking the Sierras, and looking slightly more overdressed than today's average hiker. I had always assumed that he must have been in that sort of pose when he had his epiphany – the flash of inspiration that marked the beginning of environmentalism. Or perhaps he was walking through a meadow in Yosemite, and was so overcome by the surrounding beauty that it changed his life forever, and turned him into nature's evangelist.

Well, as you might have guessed, that's not exactly how it happened. Instead, Muir's inspiration came in the months following the completion of his thousand-mile walk to the Gulf of Mexico where, instead of being overcome by nature's beauty, he was actually overcome by malaria. It was the first time that he had ever been sick. Despite his spiritual connection to the environment, here was Mother Nature trying to kill him. Up to then he'd had plenty to say about the beauty and splendor of the natural world, but now he'd gotten a real taste of its malevolent side. And that was what brought about his epiphany.

At the time, most people believed that all of God's creation was for the benefit and amusement of humanity, but Muir began to realize that the workings of nature had nothing to do with what humans might want or need.

> The world, we are told, was made especially for man – a presumption not supported by all the facts. . . . The universe would be incomplete without man; but it would also be incomplete without the smallest transmicroscopic creature that dwells beyond our conceitful eyes and knowledge.[39]

If we value nature only in terms of its utility to human society, it would mean that we could dispense with those aspects that are not pleasing to us, and preserve only what is perceived to be beneficial to humans. The natural world would end up looking like a European garden. On the contrary, Muir argued that everything in nature is connected, like an intricately woven tapestry. If you pull on a thread, then the whole thing starts to unravel.

> When we try to pick out anything by itself, we find it hitched to everything else in the universe.[40]

The only way to protect the environment, says Muir, is to preserve everything – even the bad stuff that tries to kill us, or the ugly plants and animals that nobody likes. That was the idea that gave birth to modern environmentalism.

It's hard not to notice the role that religion plays in all of this. What frustrated Muir was that people were all too ready to make up their minds about nature, having been informed, not by the facts, but by their religious outlook. He felt that the level of confidence that people had in their beliefs was unwarranted.

> They have a precise dogmatic insight of the intentions of the Creator . . . He is regarded as a civilized, law-abiding gentleman in favor either of a republican form of government or of a limited monarchy; believes in the literature and language of England; is a warm supporter of the English constitution and Sunday schools and missionary societies; and is as purely a manufactured article as any puppet of a half-penny theater.[41]

Frustrating though it may be, history suggests that this is how people usually operate. We have a tendency to be overconfident in our worldview, even when it is based entirely upon religious doctrine. Anthropologist Donald Grayson provides another example:

> At the beginning of the eighteenth century, it was widely accepted that both the human species and the earth had been created about 6000 years ago. This belief was in turn associated with the beliefs that the words of Genesis were to be read literally, and that since the earth had been created for people, an earth without people was an earth without purpose.[42]

We know, of course, that the world is a lot older than 6,000 years, although I imagine that for a few members of Congress, the eighteenth century view may still prevail. Muir discovered when you get out into the world and examine things with a critical eye, our preconceptions don't always hold true. Experience shows that the received wisdom we get from religious belief is usually little more than unsophisticated guesswork, so chances are, it's going to be wrong. That's why the Pope ended up apologizing to Galileo in 2000, and why in 2008, the Church of

England said they were sorry about how they had treated Darwin.[43] [44]Out side the United States, at least, Christians seem to have given up trying to take the Bible literally.

I am sure we could forgive our religious leaders if they decided to stop making pronouncements altogether, for fear of having to apologize later. These apologies to scientists from the Christian hierarchy are presumably intended to set things right with a more scientifically literate congregation; after all, the scientists to whom they are apologizing are long dead. These days, it's difficult to be credible if you're considered to be unscientific, so it makes sense to market Christianity as being fully compatible with the established tenets of science. But how compatible can the two really be?

Generally speaking, the language of science doesn't involve absolute certainties. When the results of an experiment are published in a scientific journal, the authors will usually draw some conclusion from the data, but they will also include other possible arguments that could lead to differing interpretations. Scientific papers are filled with error bars, degrees of uncertainty, and discussions about plausibility. Researchers are careful to distinguish between causation and association – in other words, whether the experiment shows an actual cause and effect, or if two things just appear to be related in some way. Reverse causality may also be considered. For example, suppose an experiment shows that people who have high levels of a particular chemical in their blood are more likely to have something wrong with their kidneys. The conclusion might be that the chemical causes kidney disease, whereas the real explanation might be that in people with kidney disease, the kidneys are just not doing a good job of extracting the chemical from the blood.

Scientist Jacob Bronowski explains that, even in relation to the fundamental forces of nature, the intrinsic uncertainties of quantum mechanics make it impossible for us to be definitive about reality. The world as we understand it is full of gray areas; nothing can be regarded as absolute.

> All knowledge, all information between human beings can only be exchanged within a play of tolerance. And that is true whether the exchange is in science, or in literature, or in religion, or in politics, or even in any form of thought that

aspires to dogma. . . . It is an irony of history that at the very time when this was being worked out there should rise, under Hitler in Germany and other tyrants elsewhere, a counter-conception: a principle of monstrous certainty.[45]

And yet seen through the lens of religion, things are mostly black and white – a point that both Pope John Paul II and Pope Benedict XVI have taken pains to emphasize in stressing the importance of absolute beliefs. Bronowski, however, takes a very dim view of dogmatism.

> This is the concentration camp and crematorium at Auschwitz. . . . Into this pond were flushed the ashes of some four million people. And that was not done by gas. It was done by arrogance. It was done by dogma. It was done by ignorance. When people believe that they have absolute knowledge, with no test in reality, this is how they behave.[46]

Bronowski obviously provides a very extreme and powerful example, but his point, I believe, is generally applicable. In terms of their underlying philosophies, science and religion couldn't be further apart. It's difficult to be dogmatic, and to always be right. In fact, there are those who wonder if there's really a place for this kind of absolutism in progressive society. In their book *The Population Explosion*, scientists Paul and Anne Ehrlich recall that in 1988, as the Pope reaffirmed that contraception was wrong, his supporting cast of bishops told us that there was no need for birth control since "the world's food resources theoretically could feed 40 billion people."[47] The Ehrlichs were troubled by the ambivalence of the bishops toward the suffering that could result from that much overcrowding.

> One might also ask whether feeding 40 billion people is a worthwhile goal for humanity, even if it could be reached. Is any purpose served in turning Earth, in essence, into a gigantic feedlot?[48]

One might contend that, on matters of science and public policy, religion might not be the best oracle to consult. If this were a Mel Brooks movie, I'm betting the next scene would show an assembly of robe clad atheists responding to the Catholic bishops in a harmonized reworking of the popular Edwin Starr song.

> [Religion,] huh, yeah
> What is it good for?
> Absolutely nothing.[49]

Nevertheless, religion and power have traditionally been inseparable in many parts of the world. In Ireland, for example, the authority of the Roman Catholic Church went unquestioned for most of its history. When the country became an independent republic in 1937 there was little attempt to separate church and state, particularly with respect to the education of children. The Preamble to the Irish Constitution gives the game away pretty quickly.

> In the Name of the Most Holy Trinity, from Whom is all authority and to Whom, as our final end, all actions both of men and States must be referred,
> We, the people of Éire,
> Humbly acknowledging all our obligations to our Divine Lord, Jesus Christ…[50]

And just in case there was any doubt about Ireland's allegiance to religion, the matter is explicitly put to rest in Article 44.

> The State acknowledges that the homage of public worship is due to Almighty God. It shall hold His Name in reverence, and shall respect and honour religion.[51]

In recent times, however, Ireland's respect for the Catholic Church dwindled when it was discovered that priests had been sexually abusing kids for decades, and that Church leaders knew what was going on, and made deliberate attempts to hide the details from law enforcement. Many began to question whether they should put their faith in a system that knowingly permits such atrocities within its ranks. In 2011, Irish Prime Minister Enda Kenny openly criticized the Vatican for trying to "frustrate an inquiry" into child abuse.[52] It was the first time the Irish government had ever castigated the Holy See in this manner.

> This is not Rome. . . . This is the Republic of Ireland. . . . A republic of laws, of rights and responsibilities; of proper civic order; where the delinquency and arrogance of a particular version, of a particular kind of "morality," will no longer be tolerated or ignored.[53]

It's hard to fathom why the Church would continue in its attempts to bury the scandal, since this can only make things worse from a public relations standpoint. One can only assume that these actions are the result of significant dysfunction within the organization, which I suppose is what we should expect from a secretive culture that discourages criticism. The Church asks for trust, and yet its behavior is anything but trustworthy.

Figure 10. *The Creation of Adam* – painted by Michelangelo on the ceiling of the Sistine Chapel.[54]

There's a certain time each year when Catholics are meant to subject themselves to a little austerity as a form of penance for their sins. It's called Lent, and it takes place over a six week period leading up to Easter Sunday. My dad, Jim, told me that during Lent the priests would do a lot of preaching about repentance and self-denial, and the people took heed. His father worked at a foundry, and every morning he packed himself a sandwich for lunch before he left for the day. However, when Lent came around, my grandfather would leave the brown bag behind, and would work a full day without any lunch. Jim remembers one evening, when his mom asked him to bring a message over to his brother Billy at the monastery. Uncle Billy was in training to become a Christian Brother at the time, although he later came to his senses and got married instead. Anyway, Jim arrived at the monastery to find all the brothers enjoying a dinner spread that was reminiscent of one of King Henry's banquets.

"I thought we were supposed to be cutting back?" he inquired.

"Oh, yeah," replied Billy. "We normally have four courses, but we've dropped down to three for Lent."

What then can we say about the purpose of religion? What is it good for? The obvious answer is that religion provides us with our moral compass, and keeps society on the right track. We need religion to remind God-fearing people that there's a God to fear – otherwise they might run amok. Then again, Nobel laureate Steven Weinberg reminds us that religion doesn't have a great track record on morality either. For hundreds of years, religion co-existed quite happily with slavery, says Weinberg, and when it was finally abolished – slavery, that is – it was because of the progressive changes happening in society at the time, and not because of any moral imperative raised by religion.

> As far as I can tell, the moral tone of religion benefited more from the spirit of the times than the spirit of the times benefited from religion. . . .
>
> With or without religion, good people can behave well and bad people can do evil; but for good people to do evil – that takes religion.[55]

Many public figures like to place great emphasis on what they believe, as though the mere act of believing is a virtue unto itself. It seems to me, though, that the currency of belief has been overused and devalued, especially when it is deliberately employed to supplant compassion, or chosen in place of knowledge and understanding. Beliefs about abortion, homosexuality, stem cell research, and women's rights are all championed by well-meaning, God-fearing folk; but when we start acting on those beliefs, compassion often gets thrown out the window. Also, it makes no sense to walk around believing something when the facts of the matter can be readily established. Surely it is better to know than to believe. There are far too many beliefs about global warming, for example. In the end, as Muir points out, the true nature of reality doesn't change just because we believe things to be different. Just because we say that the world is 6,000 years old, doesn't make it so.

> Our own good earth, made many a successful journey around the heavens ere man was made, and whole kingdoms of creatures enjoyed existence and returned to dust ere man appeared to claim them. After human beings have also played their part in Creation's plan, they too may disappear without any general burning or extraordinary commotion whatever.[56]

Given that religion has had to apologize for being wrong about the things that are actually verifiable, what kind of accuracy can we expect on other important questions, such as the existence of God, or life after death? Where do we stand on those things? If religious doctrine couldn't reveal the simple fact that planets orbit stars, and not the other way around, then why should we believe revelations about angels and gods and the like? In other words, there's a good chance that religious truths might not be, well, true – although conveniently enough for the theologians, we don't always have an easy way to check.

In the end, I guess it depends upon how much you want to believe, or whether you were properly indoctrinated as a child. It reminds me of *Father Ted*, a British comedy about three Irish priests living in a small community on some remote island. In the episode "Tentacles of Doom," three bishops come to visit the island, and one of them asks Fr. Dougal McGuire if he ever has any doubts about his religion.

"So Father, do you ever have any doubts about the religious life? Is your faith ever tested? Anything you've been worried about? Any doubts you've been having about any aspects of belief? Anything like that?"

"Well," replies Dougal, "you know the way God made us all, right? And he's looking down on us from heaven, and everything."

The bishop nods in acknowledgement.

"And then his son came down and saved everyone, and all that."

"Oh, yes," says the bishop.

"And when we die, we're all going to go to heaven."

"Yes, what about it?" asks the bishop.

"Well, that's the bit I have trouble with," confides Dougal.[57]

In the very first episode of the series, Ted and Dougal are at a local fair, when Dougal spots a fortune teller's tent.

"Don't waste your money on that stuff Dougal," says Ted.

"Come on Ted," insists Dougal. "You never know, there might be something in it."

"It's rubbish," declares Ted. "How could anybody believe any of that sort of nonsense?"

"Come on Ted, sure it's no more peculiar than that stuff we learned in the seminary; you know, heaven and hell and everlasting life, and all that type of thing. You're not meant to take it seriously Ted."

Of course, as a frustrated Ted then exclaimed, "[We] are so too meant to take it seriously!"[58]

Actually, according to the statistics, most of us do. And if so many of us are convinced that Earth is nothing more than a way station, might that pose a problem for environmentalists? I mean, if our final destination is Heaven, then how committed are we to saving the planet? Most western religions maintain that we're not supposed to put too much stock in the things of this world, as it is merely a staging ground for the next life. To me, that doesn't sound like we're all equally invested in Earth's future. Luckily for Earth, these religions also hold that one's place in the next life depends upon one's performance in this one, so perhaps our stewardship of the planet will be one of the factors that get taken into consideration.

How seriously should we take this God idea? It's clear that most of our normal everyday activities have been effectively secularized, and don't involve a lot of religious deliberation. In my experience, supernatural or magical thinking is not considered good business practice, and wouldn't be tolerated in the workplace. If our engineers tell me that the system isn't working, I expect them to come up with a plan to fix it. Praying over the machine is not an option, nor do I accept the explanation that an evil spell has been cast upon the software.

Any physicists that I've talked to over the years have had very little to say about God, and they tell me that this is because you can't prove whether God exists. However, on a different day, they'll tell you that science isn't about proving or disproving anything. It's about coming up with a good working hypothesis – one that seems plausible, and that fits the available evidence. Choosing a supernatural being as an explanation for the existence of the universe doesn't sound like a good working hypothesis to me. While my physicist friends seem happy to declare that religion is outside of science, I've discovered that anthropologists, sociologists and psychologists seem to have a different opinion. From their point of view, science has quite a lot to say on the subject.

The best working hypothesis from a sociological point of view is that religious ideas were manufactured in the minds of people.[i] The simplest explanation, in other words, is that the gods do not really exist, and it is for this reason that civilizations such as ours will invariably invent themselves a god. As Diane Ackerman explains, "The brain is a pattern-mad supposing machine."[59] We look for meaning and purpose in everything, and so when we see people suffer and die for no apparent reason, we are compelled to find a way to make sense of it all. Religion allows us to do this collectively, which strengthens and legitimizes our beliefs. Our belief system thus becomes woven into the strong social fabric of society. The invention of writing gave us the means to permanently preserve our religious doctrines in the form of sacred artifacts. Religion is a way to rationalize those aspects of our existence that would otherwise seem meaningless.

When I was a kid we went to church once a week to attend Catholic Mass, and overall the experience was fairly harmless. It was, after all, a nice way of meeting all of the neighbors, since in those days, pretty much everyone in the community showed up at church. Unfortunately we never really accomplished much in these weekly get-togethers. The priest told us a few stories, but they were always ones we'd heard before, and his sermons were like John Wayne movies, in that we always knew how they were going to end. They ran the same show each Sunday, and yet they were guaranteed a huge audience every time. It was really a wasted opportunity, when you think about it. Imagine if we'd had a guest speaker each week, to give us some gardening tips, or show us a slide presentation about the rainforests, for example. At least we would have learned something. What if we had used that time slot to have a weekly

---

[i] This is explored in more detail in Chapter 1.

town hall meeting about some important political issue, or some community problem that needed sorting out? That would arguably have been a better use of our time.

At least we can say that religion brings people together, which is a good thing. It provides us with a common framework, a set of values and guidelines that gives us a sense of kinship with the rest of society. As I remember it, one of the more useful functions of the Church was that it provided a conduit for folks who wanted to help out in the community, by visiting the sick, for example, or providing assistance to those in need. Among those selfless individuals were the priests themselves, and it was their work of charity and compassion that supplied, in Weinberg's words, the "moral tone" that we associate with religion.[60]

Of course on the negative side, we must also acknowledge that this powerful unifying force of religion can also serve to alienate people who are of a different persuasion. As the troublemakers of history have shown, religion is a convenient cultural marker for those interested in creating conflicts between groups.

If you've grown up in the Catholic tradition, then you'll know there's a certain ritual that's performed when you're about eleven years old. Thankfully it doesn't involve getting your foreskin chopped off, but you are expected to profess your faith and belief in all things Catholic, so that you can be "confirmed" as a full card-carrying member of the Church. This confirmation process involves being able to answer any question the bishop might throw at you, so the kids have to prepare for the ceremony by memorizing a lot of Church doctrine. If you ask any of these kids why their religion is important to them, they will likely be able to give you a well-crafted response. But given that they've only been on the planet for eleven years, how is it that they feel so strongly about being Catholics? Why not something else?

The answer, of course, has little to do with the merits of Catholicism, or the persuasive accuracy of the Catechism. In all likelihood, the kids are Catholic because their parents were Catholic, just as they would have been Jewish if their parents were Jewish. Their religion is in their upbringing, or as many will tell you – it's in their blood. Community begins with the family, and family ties are very important to all of us primates. So the fact that people are serious about their religious beliefs tells us nothing about the strength of evidence supporting those beliefs.

One of the questions that came up in our religion class at school was whether it was appropriate for us all to be brainwashed in this way. I felt a certain admiration for the kid who posed the question, as this wasn't the type of thing you would typically ask a Christian Brother. We all remained silent for a moment, to give the good brother a chance to digest the query. Then the kid offered a clarification.

"I mean, there are so many religions in the world, and they can't all be right. So how do we know that ours is the one true religion?"

The brother seemed more comfortable with this rephrasing of the question, as his face softened a little.

"Well," he said, "when you get older, you can choose a different religion if you want. But at least you'll have Catholicism to use as a benchmark; something to compare all the others against."

His response was actually well received. We all nodded, and considered the matter closed. I even heard someone say, "Yeah, that makes sense."

However, in later years, as I recalled that particular episode, I realized that the brother was telling us that we could make whatever choice we wanted, as long as it was religion. It was like Henry Ford's pronouncement that we can have whatever color we want, "so long as it's black."[61] The lesson for us kids was that it was important to get used to the practice of believing what one is told, or what is revealed by some holy book. The menu of choices did not include critical thinking.

According to Martin Luther, founder of Protestantism, any religion worth its salt should have something to say about the afterlife.

> If you believe in no future life, I would not give a mushroom for your God![62]

Luther's got a point. For me, that's the most compelling aspect of religion – the notion that we might meet our loved ones again after we die. I was ten years old when Granddad Waine died, and it was then that I became especially interested in the idea of heaven, since that was where he was supposed to have gone. So, in typical kid fashion, I thought I'd ask a few questions. But as much as the adults liked to talk about resurrection and eternal salvation, they were pretty skimpy with the details.

In the '70s and '80s, we began to see stories on television about people who were taken to the brink of death by some severe trauma, such as an accident or a major operation. After being resuscitated, some of these individuals then gave curious accounts of how they had caught a glimpse of their loved ones waiting for them in the next world, or how they found themselves floating outside their bodies, looking down on the surgeons from above. This, we were told, was proof that the afterlife existed. For me, though, it wasn't convincing enough. A far simpler and more plausible explanation would be that these are purely hallucinations, and have nothing at all to do with everlasting life.

In any case, the topic never came up for debate during any of our weekly church visits, nor did the priests ever invite any questions from the audience. As far as I could make out, there was no real evidence of the existence of heaven, and no good reason to think that people can survive death. What was more – none of the churchgoers seemed to mind. Perhaps Dougal was right. Maybe we're "not meant to take it seriously."[63]

So what do the scientists have to say about all this? As we've seen, there are plenty of scientists who are so fed up with religion that they are always ready to disparage the faithful. They have their own dedicated followers, and they usually mean well, but it's generally the case that people don't like them. They wouldn't stand a chance in a US election.

Then there are the more popular scientists; the ones who are friendly toward religion. Astrophysicist and TV presenter Neil deGrasse Tyson admits that some scientists don't mind pretending to be religious, if it helps them to sell more of their books.

> Publishers have come to learn that there is a lot of money in God, especially when the author is a scientist and when the book title includes a direct juxtaposition of scientific and religious themes. . . . While the books are not strictly religious, they encourage the reader to bring God into conversations about astrophysics.[64]

The famous physicist Stephen Hawking is a self-professed agnostic, yet in 1998, when asked if God really existed, "I do not answer God questions," was his response.[65] He came clean in 2011, however, and told everyone that the answer was no.⁶⁶

When I was a boy, I wondered if the scientific community would ever find proof of the existence of God. There was, however, not much chance of that happening. To start with, as my religious mentors explained, God would never reveal himself in that way. It would make things too easy for us. Also, the truth is that most of the big names in science don't care much for God, despite whatever titles they use on their dust jackets. They much prefer the true nature of reality to the mystical mumbo-jumbo of religion. Muir seemed to feel the same way. He believed that getting out into the wilderness was like returning home. There was something familiar about nature, and it allowed him to find his true self. Over time, this new kind of spiritualism gradually replaced the traditional religious notions he had grown up with.

> I never tried to abandon creeds or code of civilization; they went away of their own accord, melting and evaporating noiselessly without any effort and without leaving any consciousness of loss.⁶⁷

Astronomer Carl Sagan often talked about the deep sense of wonder that comes from understanding our place in the cosmos, and how we are all made from star stuff. For him, scientific truth was much more compelling, and more spiritually uplifting than the fabricated stories of religion. In debunking the creationist myths, here's what he had to say about evolution.

> A Designer is a natural, appealing and altogether human explanation of the biological world. But . . . there is another way, equally appealing, equally human, and far more compelling: natural selection, which makes the music of life more beautiful as the aeons pass.⁶⁸

We can all feel that spiritual connection to nature, I'll not deny it. But can we really say that this tingling of scientific wonderment is as compelling as the idea of a father creator, or the prospect of being reunited with family and friends who have died? Sorry Sagan, but I don't think so. Religious mumbo jumbo definitely has the edge. In fact, religion beats the pants off science in many respects. Take Christmas, for example. That's a lot more fun than any science fair I've been to recently. And it must be said that the story of Jesus is actually not that bad. It's a much more interesting read than most of the scientific papers I've come across. However, as Luther says, the most compelling aspect of religion is the promise of a solution to death. Let's face it, everlasting life is a lot better than anything science has to offer. For one thing, it gives Sagan the opportunity to argue the point with us later, over a few heavenly beers.

To be fair to Sagan, it's not as though he was in a hurry to die. Like the rest of us, he would have welcomed the chance to stick around a bit longer given the chance.

> I would love to believe that when I die I will live again, that some thinking, feeling, remembering part of me will continue. . . . I want to grow really old with my wife, Annie, whom I dearly love. I want to see my younger children grow up and to play a role in their character and intellectual development. I want to meet still unconceived grandchildren. There are scientific problems whose outcomes I long to witness.[69]

For those who have a great passion for life, death is always an unwelcome intrusion. We'd probably never have gotten rid of Benjamin Franklin if the choice had been left up to him. The problem with all of this, though, is that it isn't real, or at least it wasn't real enough for Sagan.

> But as much as I want to believe that, and despite the ancient and worldwide cultural traditions that assert an afterlife, I know of nothing to suggest that it is more than wishful thinking. . . . There is no reason to deceive ourselves with pretty stories for which there's little good evidence. Far better, it seems to me, in our vulnerability, is to look Death in the eye and to be grateful every day for the brief but magnificent opportunity that life provides.[70]

Most Americans are happy to take things on faith, and don't seem to be too bothered about the lack of evidence. For some of us, though, the faith approach doesn't work. We just cannot see the point in trying to believe in something that's completely implausible, and as the nineteenth century agnostic Robert Ingersoll observed, everlasting life definitely falls into that category.

> The only evidence, so far as I know, about another life is, first, that we have no evidence; and secondly, that we are rather sorry that we have not, and wish we had.[71]

Nevertheless, if you ask me, it's a real shame that God doesn't exist. I was looking forward to going to heaven. It also means that we're out of options as far as resurrection is concerned. As Gilgamesh witnessed after the death of his friend Enkidu, once you're dead, you're dead.

> How can my mind have any rest?
> My beloved friend has turned into clay.[72]

Unfortunately the same applies to Granddad. He now exists only in the thoughts and memories of those who benefited from having known him when he was alive.

The conclusion, I guess, is that it's not good to be dead, but what does it mean for the living? For many of us, as the gods begin to fade, along with them go our dreams of heaven. So where does that leave us? I suppose that it leaves us here, on the good earth, where we started in the first place. Contrary to what we learned in church, it's this world that's important – not the next. In fact, there is no other world waiting for us, so it might behoove us to show more appreciation for the one we have. It is, after all, the only home that humanity has ever known, so if it's heaven we're after, we'll have to find it on Earth.

It also means that there are no second chances. If we only get one shot at life, then we need to be extra careful not to waste it. For me, this underscores how precious life is. We don't have the luxury of consulting the scriptures, or the prophets, so that when things start to go pear shaped we can just throw up our hands and say that it's all part of God's grand plan. If there's no God, then there's no plan.

Maybe Muir was onto something when he talked about going out into the wilderness to find himself. Perhaps it's time for us to do the same, to forget about heaven and come back down to Earth; to return home, as Muir would say, so that we can find that sense of the spiritual that comes from knowing who we are, and where we belong.

> We shall not cease from exploration
> And the end of all our exploring
> Will be to arrive where we started
> And know the place for the first time.[73]
>
> - T.S. Eliot, "Little Gidding" (Four Quartets)

8

⌘

The Age of Death

Every so often I'd find my dad, Jim, standing outside in his back yard, gazing up at the stars.

"There's got to be some point to all of this," he used to say, with his eyes fixed on the heavens. "There must be an answer out there."

Being just a kid, it wasn't always clear to me what the questions were – but at least I knew the answers were to be found in outer space. There was one occasion when, after a few beers, he told me that the answer had to do with magnets. But most of the time it was outer space.

When I began watching *Star Trek*, I realized that my dad was onto something. By a happy coincidence the captain of the Enterprise was also called Jim, and he exhibited the same pragmatic, no-nonsense approach to life that I admired in my dad. True enough, outer space seemed to have all the answers to humanity's problems, and so in my mind it was imperative for us to get out and start exploring the galaxy as soon as humanly possible. There were new worlds to explore, aliens to meet; aliens that, by the way, would have vast encyclopedias of knowledge, and would be able to cure all of our diseases, and might even be able to offer us immortality. In the *Star Trek* universe, there were planets where alien technology could conjure up anything you desired; there were energy fields that could turn humans into gods, and, as a bonus feature, it seemed relatively easy to travel backward in time, if we wanted to do that. Besides which, as Jim Kirk said, space was the final frontier, and the idea that we might just ignore a frontier was certainly contrary to my way of thinking.

I wasn't stupid though, and I knew that *Star Trek* was just a TV show. Those aliens might not actually exist. However, in the 1980s, a new advocate for space travel showed up on television. He wasn't a starship captain, and his name wasn't Jim, but he had a couple of things going for him. As you may have guessed, I'm talking about the astronomer Carl Sagan, presenter of the popular PBS series *Cosmos*. Sagan had the advantage of being an actual scientist, which made him slightly more credible than the fictional Mr. Spock. *Cosmos* tried to show the universe as it really was, rather than how science fiction writers pictured it to be, which admittedly sounds like a recipe for a comparatively boring production. The success of the series, though, was in demonstrating that the real universe is actually more awe-inspiring than the fictional one. The truth, says Sagan, is more incredible than anything we could have imagined, and it is all the more compelling because it is real.

Cosmos was shown in more than sixty countries worldwide with an estimated viewing audience of 500 million people who all learned that the cosmos doesn't really feature a lot of what we've come to expect from *Star Trek*. There are no aliens with pointy ears, nor is there any immediate prospect of us gallivanting across the galaxy in a fleet of starships. Sagan did, however, point out that human beings have a real, honest-to-goodness connection with the cosmos, in that the very atoms that make up our bodies were manufactured in the interiors of the stars. We are, in other words, "made of star-stuff."[1]

He also reminded us that the same natural laws that govern our lives here on Earth are equally applicable throughout the cosmos, which means that the same conditions that allowed life to flourish on this planet have likely prevailed on other worlds as well. Life may be abundant in the universe. Sagan hinted at the possibility of intelligent life, which I guess provides some corroboration for *Star Trek* on at least a couple of key points; the first being that it is possible, if not extremely probable, that there lots of life bearing planets out there, and secondly, that it is entirely possible that some of those planets might be home to advanced alien civilizations.

Now before we get too excited, let me quickly point out that *Star Trek* is completely off the mark in suggesting that the galaxy might be filled with lots of humanoid species, all having reached approximately the same level of technological development. That's the basic premise for many exciting episodes, and it's almost certain to be complete balderdash. If you could use a time machine to travel back to some random point in Earth's history, odds are that you would arrive at a time long before humans inhabited the planet. You'd find plenty of life, but nobody to talk to.

If evolutionary biologist Ernst Mayr is correct, the situation will be no different with respect to life elsewhere in the galaxy; we'll find lots of strange looking animals and plants, but no human-like intelligence. Technologically speaking, humans are very young; our civilization is only a few thousand years old, which is not much when compared to the age of the universe. So if we do ever stumble upon intelligent aliens, supposing they exist, they will most likely reside somewhere toward the god-like end of the technology spectrum, from our point of view. It would be extremely unlikely for us to happen upon a species that is exactly level with humanity on the timeline of progress. That means we can rule out most of the *Star Trek* cultures, including the Klingons and the Vulcans. The reality of extraterrestrial life may be less like Gene Roddenberry's universe, and more like the one portrayed in John Carpenter's *Dark Star*.

> Remember when Commander Powell found that ninety-nine plus probability of intelligent life in the Magellanic Cloud? . . . Remember what we found? A damn mindless vegetable; looked like a limp balloon. Fourteen light years for a vegetable that went "Squawk" . . . remember that? . . . Don't give me any of that intelligent life stuff.[2]

The biggest element that we are missing in order to achieve our *Star Trek* vision is warp drive. That's the fictional method of propulsion that the U.S.S. Enterprise employs to traverse the huge distances between the stars. Although my high school science teacher gave *Star Trek* some kudos for suggesting that anti-matter might be a viable source of energy for the ship's engines, it needs to be said that, so far at least, science has provided no evidence that supports the concept of warp drive. However convincing the technobabble might be, and despite occasional claims to the contrary by overzealous physicists, this futuristic technology rests entirely in the realm of science fiction, and is likely to remain there for the foreseeable future – perhaps even indefinitely.

Without some advanced form of transportation it could take tens of thousands of years just to get to the nearest star systems, which means that we can forget about interstellar space travel for the time being. Now that we've killed Roddenberry's dream, where do we go from here? Sagan is adamant that we should continue with the reconnaissance of the solar system for now. He admits that this was, in fact, the primary motivation for making the *Cosmos* series in the first place – to encourage people to take an interest in the robotic exploration of the solar system. Hopefully we would then be more inclined to urge Congress to keep spending money on space probes.

At around the same time *Cosmos* was launched, Sagan got together with Bruce Murray and Lou Friedman to set up The Planetary Society, a Pasadena based non-profit that would serve as a lobbying force for space exploration and the search for extraterrestrial intelligence. I was still fairly convinced that the answers were in outer space, so I enthusiastically agreed to become the society's ambassador in Ireland, and spent the next eleven years working to educate the public about planetary science, and the importance of continuing the space program. We told everyone that space exploration was a worthy goal for humanity; that by exploring other worlds we could learn about our own, and that by finding life elsewhere, we could learn about ourselves and our origins.

Eventually, says Sagan, if we are to ensure our survival, we'll need to become a multi planet species; so we might as well start working on that now.[3] Besides, he says, exploration is in our blood; we are destined to head out into the cosmos.

> Some part of our being knows this is from where we came. We long to return.[4]

Sagan also suggested collaborating with the Soviet Union on a trip to Mars as a strategy to build international goodwill and diffuse some of the cold war tension between the superpowers. He argued that such collaboration provided a way for us to redirect our missile technology toward peaceful ends, and to keep our military scientists on the payroll by having them beat a few more swords into plowshares.[5] That was a good argument, but it's lost a lot of its potency now that the Soviet Union no longer exists. Becoming a multi-planet species probably makes sense in the long term, but as Granddad and I surmised back in the seventies, we have lots of time before the Sun runs out of fuel, so what's the hurry?

Physicist Neil deGrasse Tyson worries about what might happen if we figure out how to keep people alive forever. The world, he says, would rapidly fill up with people, which would overwhelm the biosphere. How can we avoid such an environmental catastrophe? The answer, he tells us, is outer space. We can start shipping people out there to make more room here on Earth.[6] This is another fairly common theme in science fiction; the idea that if we make a mess of Earth, we can still survive as a species, as long as we've established some colonies out in space. That's what almost happened in Ray Bradbury's *The Martian Chronicles*. However, when word reached Mars of an imminent global conflict on Earth, the colonists stupidly returned to their home world and were annihilated with the rest of us.

Steven Spielberg's *Terra Nova* was based on a similar premise. The series begins by showing us a future in which industrialization has made Earth almost uninhabitable. People have to wear special respirators so that they can breathe the heavily polluted air, while wealthy citizens build domed cities to protect themselves from the harsh environment. As resources run low, and all hope is lost, a few fortunate pilgrims manage to escape these desperate circumstances by establishing a new colony, but this time it doesn't involve outer space. Instead, they find a portal that lets them travel back to the time of the dinosaurs, 85 million years in the past. This prehistoric refuge gives humanity a second chance.

That's all well and good, but since there are no portals to the past, the time travel option is out. If we're serious about getting off the planet, then the space buffs will tell you that Mars is probably the best bet. That's because you can find resources on Mars, notably water, a basic human necessity that could also be used to grow plants and manufacture fuel. However, speaking as a longtime advocate for space travel, I'd have to admit that we often oversell Mars as a destination. The truth is, it's a very bleak place; almost as desolate as the Moon.

Unlike Earth, Mars doesn't have the protection of a thick blanket of air, or a strong global magnetic field, nor is there any ozone layer to speak of. This means that any permanent Martian settlement would need to be buried under at least a couple of meters of soil to shield the inhabitants from dangerous radiation, particularly during solar flares or solar particle events (SPEs). The thin atmosphere should provide some defense against incoming micrometeoroids, although you're still going to need a pressurized spacesuit to survive on the surface, and it might not be a bad idea to take some extra precautions during meteor showers. The air pressure on Mars is miniscule, and it's also extremely cold, so that any liquid water on the surface would instantly either freeze or boil away. The ubiquitous Martian soil is likely to contain toxins, and since astronauts would get covered in the stuff, they would probably need to spend a good chunk of each day just cleaning the dust off their suits and equipment, so that it doesn't find its way into their lungs.

Figure 11. Star Trek's Lieutenant Uhura, played by actor Nichelle Nichols.[7]

The Moon is a lot closer than Mars, and yet it's still a hugely expensive exercise to send even small groups of people there and return them safely to Earth, which is why it hasn't been attempted in the last four decades. So what would be the motivation to send people to Mars? It would be an exciting adventure for sure, and it would accelerate the development of new technology, but those are hardly good enough reasons for such a costly endeavor.

Cost is not the only factor we'd need to worry about. Far more important would be the question of safety, an area of concern that has been largely swept under the rug by science fiction. Interplanetary travel would be an incredibly risky proposition for a human being, and the experience we gained from going to the Moon does little to mitigate the risk. Perhaps the most troublesome aspect of a Mars mission is that once the space travelers are no longer covered by the safe shelter of Earth's atmosphere and magnetic shield, they will be bathed in dangerous radiation for the duration of the trip.

The health risk posed by radiation is often understated by those who would send human explorers to the planets, since the reality is that astronauts on an interplanetary trajectory would have virtually no protection against cosmic rays and energetic particles in the solar wind. Unpredictable SPEs could increase the amount of radiation to lethal levels. This is less of a problem on lunar, or Earth orbit missions, because Earth's magnetosphere still affords some protection, and because astronauts can quickly abort the mission and return home, but once outside the Earth-Moon system we currently have no way to provide astronauts with adequate shielding.

Bear in mind that, whereas it takes only a few days to get to the Moon and back, a Mars trip would be measured in years. That's a long time for the human body to be exposed to hazardous radiation. Once they reached Mars, the intrepid explorers would be so far away from Earth that the travel time for a one-way radio signal could be as much as twenty minutes, making it impossible for them to have a normal interactive conversation with controllers back home. They would be completely isolated from the rest of humanity.

Astronauts would also suffer from other health effects, including muscle atrophy and bone loss due to the prolonged microgravity, or weightlessness. The more you think about it, the more you begin to question the whole concept of a human mission to Mars. If the trip doesn't kill them, they'll probably die soon afterward from cancer. There'd be no shortage of thrill seeking astronauts lining up to accept the challenge, but would such a mission be worth the risk?

The reality may be that humans are simply ill-equipped to deal with the hazardous environment beyond Earth. In that respect, we are not alone, as the same applies to all life forms. Life is everywhere on Earth, even in the most hostile and unpleasant places. On Mars, however, we haven't yet been able to find a single microbe on the planet; the environment appears to be completely sterile. From the point of view of life in the solar system, clearly there's no place like Earth.

Given this reality, is there really a case to be made for sending humans elsewhere? Does the adventure outweigh the risk? Tyson acknowledges that outer space is unsuitable for human life, but he suggests that we can remedy this by altering our bodies. Through the use of genetic reprogramming and artificial body parts, Tyson believes that we can engineer a new type of human, specially adapted to live on other planets. And he believes that this will all be possible within our lifetimes.

I can accept that we're well on the way to being able to custom build new organs, so that nobody will have to die while on the waiting list for a new heart or a kidney. It sounds, however, as though Tyson is suggesting that in this same timeframe we'll be able to completely replace an entire human body, brain and everything, with a new type of body that is specially tailored to survive in outer space. Should we take this suggestion seriously? Is the concept even plausible?

To start with, maybe we should consider the kind of body that would be best suited for survival in a lifeless, airless environment. Maybe we'd end up with something similar to the robotic probes that were designed and built to survive on the Martian surface. To Tyson's point, a body like that would be very different indeed. We might have wheels and electric motors instead of legs, and we'd probably want to dispense with the whole notion of a cardiovascular system. Maybe we'd also want to be solar powered, or perhaps we'd use radioisotope-based thermoelectric generators instead. One critical requirement is that the new body would have to contain an artificial brain with the same causal powers as a human brain, which could, in other words, serve as a host for a human mind. Could this be what Tyson has in mind?

However plausible this might sound to some people, there is simply no evidence suggesting that anything like this is going to happen in the near future. I can only assume that Tyson was just kidding around, and that his suggestion wasn't supposed to be taken seriously. More to the point, I'm not sure I like the idea of having to live inside a robotic body on some desolate world. Granted it would be a convenient way to do some exploring, as long as I could be placed back into my own body at the end of the mission. But if Tyson is telling us that we'll have to live out the rest of our days looking like monsters, and trundling around the Martian deserts, then I'd have to say that this whole space travel business is starting to look a lot less attractive. Even if we accept Sagan's logic that we shouldn't have all of humanity's eggs in one basket, so to speak, can we honestly foresee a scenario in which Mars becomes a second chance for the species?

Compared to Mars, the dry valleys of Antarctica are nothing short of paradise. So if we are really serious about wanting to live on Mars, why aren't there hordes of us setting up camp in Antarctica? The cold fact of the matter is that, compared to Planet Earth, outer space would be a dreadful place to live. As much as I am still a big fan of space exploration, I think I've reached the conclusion that outer space does not have all the answers. As President Obama might say, my position has evolved. On the topic of going to Mars, it stands to reason that in the distant future we're sure to find ways to better protect humans from the hazards of space travel. Wouldn't it make sense to just stick with the robotic exploration of the solar system for now? Then maybe in a century or so, we can reevaluate our options for sending humans to the planets.

We should not delude ourselves into thinking that our home planet is in any way dispensable; that if we mess things up on Earth, we might start over again on some distant world. What we need to understand is that, no matter how big a mess we make, Earth will always be far more hospitable than any other place in the solar system. If we want to make a future for ourselves, this is where it must happen. By the time we are ready to move out into the galaxy, we will need to have solved the problems that threaten our survival here on Earth.

For Tyson, though, the prospect of being able to fabricate new body parts is not just a useful way to conduct space exploration. It may also, he says, be the key to immortality. His optimism is likely due to the incredible progress that scientists are making in tissue regeneration and organ replacement. The expectation is that we'll soon start growing replacement organs using organic scaffolds, or special printers that spew out cells instead of ink. If you need a new heart, for example, we can just grow a replacement, so that you don't have to wait for a donor heart to become available. Since the new heart is built from your own cells, the body won't reject it as foreign, and since it's a brand new heart, it should last much longer than a used or pre-owned model.

These new regenerative techniques are quickly making the transition from fiction to reality, which brings hope that many lives will be saved over the coming decades. But could it also mean an end to the age of death? Tyson is not alone in making that particular proclamation. The argument seems to be that as we get older and our bodies begin to wear out, we can replace the failing components with brand new parts, just as we would for an old automobile, and we'll be good for another hundred thousand miles.

However, Tyson isn't very clear on how this is supposed to work in practice. How does getting a new kidney make it easier for me to become immortal? It should keep me alive a little longer, but why would I be expected to live beyond the maximum human lifespan? Should I start to get my other organs replaced as well, even though they appear to be in good working order? What about my brain? And how do we address the fact that my body is designed to fall apart as I get older? I imagine I'm still going to die, with or without a couple of new organs, unless the suggestion is that we somehow figure out how to reprogram my entire biological system to reverse the aging process. That might do the trick, but how exactly would we do that?

This is usually the point where the futurists step in, and start talking about nanobots, hoping that nobody notices the fact that we've just stepped back into the world of science fiction. The concept here is that we could build swarms of tiny robots that would swim around inside the human body and repair all the damage caused by aging. Assuming that this is plausible, which it probably isn't, and if such nanobots existed, which they don't, it seems to me that it would be very difficult to program the little critters so that they don't cause more problems than they solve.

The human body is a very complicated machine, and biochemists are still struggling to understand how everything works at the cellular level, and to unravel all of the bio-molecular happenings governed by our genomes, epigenomes and proteomes. If we decide to mess around with our biological programming, there are lots of options available to us. We can alter our genomes, modify the workings of our immune system, interfere with gene expression or protein synthesis, disable or enhance particular signaling pathways, reprogram our cells with new instructions, or we could introduce improved synthetic components. We could even augment our natural abilities with advanced prostheses, or bionics.

As with any reengineering project, a good first step is to make sure you have a handle on the normal functioning of the components you are trying to repair, or replace. That way you can protect against unintended consequences. Whether we like it or not, one thing that's clear with respect to the normal functioning of life is that death is a foregone conclusion. Within each of us the process of life involves a great deal of death and destruction. Cells are constantly dying as we get older, and with the loss of muscle fibers we become noticeably weaker.[8] Our life support systems suffer varying degrees of degradation through the natural course of events, and our bodies seem happy to accept this cumulative damage as the inevitable result of growing old.

We can easily recognize the signs of aging from outward appearances, such as gray hair, wrinkles, blemishes, and a loss of skin elasticity, but we also know that the process is more than skin deep. Each of us faces a complete system-wide degeneration that brings with it a whole host of additional problems including cardiovascular disease, osteoporosis, muscle fiber loss, weakening veins and arteries, dementia, cancer, arthritis, cataracts, and a gradual loss of vision and hearing.

The energy of life comes from the mitochondria, tiny power plants that reside in the interiors of our cells. Like little factories, the mitochondria take in raw materials and produce usable energy in the form of a molecule called adenosine triphosphate (ATP). Regrettably, these factories don't appear to have been designed according to OSHA standards, particularly with regard to how all the toxic waste is handled. During the manufacturing process, the mitochondria produce toxins known as free radicals which are then set loose to wreak havoc on the surrounding tissue. As we get older, our cells become riddled with damage as a result of all the free radical activity, explains science writer Kathryn Brown. "Our tired cells get less efficient at repelling free radicals and mopping up oxidative messes, and the damage accumulates. We begin to rust from the inside out."[9]

Scientists S. Jay Olshansky and Bruce Carnes note that, although the body is a wonderful biological machine, it wasn't really designed to perform well indefinitely. For example, they say, when our skin gets punctured, the body tightens the blood vessels so that we don't lose too much blood. Unfortunately as we age, this vasoconstriction process makes us more susceptible to problems such as "hypertension and stroke." Olshansky and Carnes also point out that in the human eye, the complicated way in which the optic nerve is arranged "increases the risk of a detached retina."[10] Ralf Dahm from the Max Planck Institute in Germany tells us that the lenses in our eyes degrade over time, leading to cataracts in our later years, and yet the cells do not have the means to rebuild or repair themselves.[11] Clearly they're not supposed to last forever. The accumulation of sugars in the body causes our tissues to become rigid over time, says science writer Lisa Melton, resulting in "thickened arteries, stiff joints, feeble muscles and failing organs."[12]

Our natural reaction might be to view such deficiencies as design flaws, and given the chance we'd probably send Mother Nature back to the drawing board to rework the blueprints. Nature, however, takes a different view. Unluckily for us, we were designed by natural selection, a process that had little interest in keeping us alive any longer than was necessary for us to produce a few successful offspring. In essence, as we get older, it is as though our bodies are bent on self-annihilation, and we are now in the final stages of the countdown. We are in a sense, designed for obsolescence.

If we are not happy with how all of this is supposed to play out, what can we do to change the outcome? Is it possible to get Nature's finger off the self-destruct button through a strategy of repairing, replacing or regenerating body parts? If the goal is to end up with a body that can live for some indeterminate amount of time, than suffice it to say we've got some fundamental problems with the existing model. Our bodies are a lot more complicated than automobiles, and they weren't designed with pluggable components in mind. To remove death from the equation, we're looking at a serious reengineering effort. Making someone immortal certainly qualifies as a non-trivial exercise, irrespective of whether we use tiny robots or specially engineered viruses. Consider, for example, the amount of effort involved in researching the use of gene therapy techniques in the fight against cancer. This is an indication of how difficult it can be just to restore normal functioning in the body. How much more difficult would it be to program the body to do something abnormal, such as not dying?

If we're determined to go down this road, wouldn't it make more sense just to abandon the human body, and instead design a new one from scratch, perhaps with more durable synthetic parts? Then there's the question of whether any of this makes sense. I don't mean to be facetious, but what if we go to all this trouble and expense to regenerate and replace your body parts, and reprogram your system to reverse the aging process, and then you walk out into the street and get hit by a truck? That would be a bit of a setback wouldn't it?

Curiously, Tyson seems to think that such elaborate reengineering will be child's play within a few decades. Immortality will be easy, he says, compared with the challenge of developing the heavy lift vehicles to launch the newly engineered humans to their new homes in outer space. Now how could anyone find fault with that reasoning? I suppose it just goes to prove that physicists have their own peculiar way of looking at the world. As Joe Biden's mother would say, "God love him."

We know that we can build heavy lift rockets, but another thing we know is that we currently don't have the means to extend human lives indefinitely. Even if we were to cure all cancers tomorrow, says Leonard Hayflick from UCSF, it would still only add about three years to the average life expectancy.[13] Fewer people would die prematurely, but it would make little difference to the maximum life span. If your ambition is to be the first human to reach two hundred, then you've got your work cut out for you. So far, the only technique that shows any real promise is something called caloric restriction.

In the 1960s, physician Roy Walford demonstrated that he could get laboratory mice to live longer by feeding them less. By reducing the amount of calories in an mouse's diet, while still making sure that it receives all the nutrients it needs, you can dramatically increase the animal's chances of living way beyond its natural life span. Scientists believe that the food shortage creates a kind of beneficial stress that causes the body to switch into a survival mode which somehow slows down the aging process. Cells throughout the body respond to the stress by invoking a number of strategies to cut back on non-essential activities and make more efficient use of internal resources. One such cellular strategy is to activate various cleanup and scavenging routines that normally come into play through a cellular housekeeping process called autophagy.[14] Specialized agents called heat shock proteins (HSPs) are also called into action to help streamline operations. The HSPs seem to have a generally positive effect on cellular wellbeing, and also help to prime the immune system against rogue cells that have turned cancerous.[15]

Although we don't yet know for sure if the technique will work on humans, there's every reason to suspect that it will. Walford was so impressed with the results that he devised a new diet for himself and began restricting his own caloric intake. He made his recipes available on the Web for anyone else who might be interested in living longer. When he met with Alan Alda in 2001, Walford reckoned that with this new approach he might live to be 110.[16] Sadly, he died three years later from Lou Gehrig's disease, at the age of seventy nine.

So the upshot is that there's a good chance we can all live to limits of old age, as long as we're willing to walk around half starved. Alda was a bit taken aback when Walford told him that if he wanted to reach 110, he'd have to reduce his weight by about twenty percent.

"I'd be a bean pole, without the beans," he quipped.[17]

For most of us, I doubt that Walford's method is what we are looking for. The psychological effects of being hungry all the time would mean that in addition to being very old, we'd also be very grumpy. Besides, as we saw in the case of Walford himself, there's no guarantee that it would work. As I said before, you might get hit by a truck. Perhaps, in time, scientists will develop a drug that will allow us to achieve the same longevity effects without having to starve ourselves. That would be better, I think.

Why is it that some people just naturally seem to live longer than the rest of us? Is it simply because random chance operates in their favor, or do they start out with some kind of inbred advantage? Bradley Willcox and his team at the University of Hawaii decided to find out, so they surveyed a bunch of healthy centenarians to see if there was anything they might have in common, apart than being very old, that is. The scientists discovered that these long lived individuals were more likely to possess variants of a metabolism gene called FOXO3A, which is believed to play a role in reducing oxidative stress. If you have the right kind of genetic variation, there's a greater chance that you will not only live to be over a hundred, but you'll also be healthier to boot.[18] In a similar study conducted by Nir Barzilai at the Albert Einstein College of

Medicine in New York, centenarians were found to possess particularly high levels of HDL, also known as the good cholesterol.[19] These studies seem to confirm that which we might already have guessed – that if you want to live to the extreme limits of old age, your genetic heritage matters.

Unfortunately for the rest of us we're not yet at the point where we can just change our genetic code on the fly, and reprogram ourselves to live longer. From what I can tell, we don't even have a straightforward way to increase the level of HDL in our blood. Apparently mine is below normal, which probably means that my chances of becoming a centenarian are not so hot. Even if we could somehow manage to give everybody the same genetic advantages as the most long lived humans, it still might not have a pronounced effect on the maximum life span. I wouldn't say no to a couple of extra decades, but is this really the most that we can expect, given all the yarns that scientists are spinning about living forever?

For the moment, I think the average self-respecting scientist would say yes – that's the most we can expect. If we want to do better than that, and start extending our warranties beyond the current limit, then I think it's fair to say that we're going to need a more satisfactory solution to the death conundrum. We could try to combat the aging process by fixing the various health issues as they arise, and by continuously applying upgrades and patches to keep our bodies from falling into disrepair, but this approach isn't going to be pleasant for anyone. As people start living longer, age related diseases will start popping up all over the place. Like a game of *Whac-a-Mole*, it will be more than we can handle to keep them under control.

Ideally if we're going to keep people from dying, we also want to make sure that they can continue to have a normal, healthy existence. The last thing we want is to stick around long enough to be able to see what it's like to decompose. This reminds me of an episode from *Gulliver's Travels*, involving a race of immortals called the Struldbruggs. Naturally Gulliver is interested in discovering the secrets of longevity, but his hopes are dashed when he discovers that the Struldbruggs somehow managed to become immortal while still remaining subject to the effects of aging. As they get older, their bodies continue to deteriorate, and yet they never die. They are consigned to an eternity of suffering in an ever worsening state of decrepitude and senility.

Historian Gerald Gruman mentions a similar tale about an ancient Greek who, like the Struldbruggs, "was granted immortality but not eternal youth."[20] Gruman explains that in the ancient world the idea of progress hadn't caught on, so there was no reason for anyone to want to live far into the future, since they would find it to be no different than the present.[21]

If you're like me, then you'll have noticed that there are some rather obvious flaws in the concept that's being presented here. I mean, to begin with, how can you be said to have immortality if your mind is gradually fading away into dementia? Wouldn't we at some point be safe in declaring that you no longer exist? In any case, how could you live forever if your body continued to fall apart? If the integrity of your biological systems is irreparably compromised, then in what sense are you still alive?

I know, I know, I'm not supposed to take it seriously. Actually, according to Roberta Borkat, the whole idea was intended as a satire on the secularist view of progress. Jonathan Swift was trying to make the point that there was, in fact, no point in attempting to improve the human condition through the use of medical technology.[22]

It sounds a little like what Elizabeth Kübler-Ross was saying in the eighties when she complained about shameful way in which terminal patients were being treated. Kübler-Ross felt that by focusing exclusively on life prolongation, doctors were guilty of ignoring the extra suffering this might cause for patients. Instead, she argued for a more compassionate approach, one that allowed patients to live out their final days with some comfort and dignity. It was this mindset that gave birth to the hospice movement in the US.

Since we do not have a way to stay alive ad infinitum, the best we can do for now is to put some of our thoughts in writing, just as I am doing at the minute, so that some small part of our personality might survive after we're gone. In 2012, NBC news anchor Brian Williams told us that we were about to lose something very precious, and that there was nothing anybody could do about it. He was talking about all the people who were alive during the Second World War, the "greatest generation". Someday very soon, he said, they'll all be dead. The rest of the news report then went on to describe how the few that remain are being encouraged to get their memoirs written, before it's too late.[23]

Jason Leigh at the University of Illinois, Chicago is taking this memoir concept to the next level. Instead of getting people to put their thoughts in writing, Leigh's approach is to encapsulate the information in the form of a computer avatar, an animated character that behaves just like the person it's supposed to represent.

The eventual goal would be to create something similar to what we saw in the 1978 version of *Superman*. As you may remember, Superman's dad, Jor-El, is killed at the start of the movie, when his home planet Krypton explodes. Before he dies, however, Jor-El is able to dump his memoirs onto a set of crystals that he then bequeaths to his son. Later in the movie, Superman is able to use these crystals to consult with an interactive simulation of his dad. The virtual Jor-El appears to have access to all of Jor-El's knowledge and memories, and it even exhibits the same rather somber personality.

Leigh is thinking that it would be nice if we all had the luxury of being able to summon up computerized versions of our own dear departed, so that we can seek their advice on matters of personal importance.[24] I must say, though, that I can't see my dad having the patience for that sort of thing. It's difficult enough to get him to write anything down on paper; where would he find the time to program an avatar?

In any event, this avatar business isn't going to be enough to satisfy everybody. As Woody Allen says, "I don't want to achieve immortality through my work. I want to achieve it through not dying."[25]

While we could create an avatar that looks like Woody Allen, and behaves like Woody Allen, it still wouldn't have a sentient mind. I know what you're thinking, but let's not go there. My point is that it wouldn't feel like something to be the avatar. Presumably if we wanted to satisfy Allen's criteria for immortality, we'd need to know a lot more about what's going on inside his head. As uncomfortable as that sounds, the reality is that if we want to preserve Allen, then we must find a way to preserve his brain, because that's where all the important stuff happens.

We've talked about the fact that if you lose an arm, or a leg, or a kidney for that matter, that pretty soon we'll have the technology to grow a new replacement. Whether the replacements are organic, or bionic, one way or another we'll have you back on your artificial appendages in no time. The one part of the body that we can't replace is the brain, because that's the part that you need in order to be you. Christopher Reeve, the actor who played Superman, was injured in a horse-riding accident that left him paralyzed from the neck down. Despite being completely debilitated, there was no question that Reeve was still very much alive, and we know this because his brain was able to tell us.

The brain, without question, is the most important part of the body, because we're almost certain that the brain is the seat of consciousness. I say almost, because there is the argument that, since we don't yet fully understand what causes consciousness, maybe we shouldn't be too emphatic about where it resides. Some have even suggested that the spinal cord may exhibit some form of consciousness. However, as far as neuroscientist Christof Koch is concerned, such arguments are not very useful. In his view, consciousness requires "reasonably complex, non-stereotypical behavior that involves online, dynamic storage of information," and so far at least, there is no evidence that this is happening in the spinal cord.[26]

As it stands today, very little brain information is preserved. Apart from our memoirs, emails and video recordings, all of our thoughts and memories are lost forever when we die. That's not to say that humans don't like to preserve things. You only have to look at some of the garages in our neighborhood to get a sense of how reluctant people are to throw stuff away, although as my wife will tell you, I tend toward the opposite extreme. I'm always getting rid of what I consider to be junk, and I'll acknowledge that it hasn't always worked out well. Now that our lives have become increasingly digitized, we've gotten much better at preserving information. We used to keep all of our household records in a couple of old shoe boxes, but now it's all automatically cataloged, encrypted and backed up to a secure online storage site, available through the Web whenever we need it.

You'd be amazed at the precautions that businesses take in protecting their corporate data. In the case of most large corporations, all of their customer records and financial transactions are stored offsite at secure locations. Several years ago, I sat in on a presentation that described some of the security measures that were in force at one of these sites. The view from the outside showed a high perimeter wall that was reinforced to withstand heavy impacts or explosions. All of the building materials were bullet proof, with extra Kevlar shielding in places. There were armed guards and security cameras everywhere. Businesses generally utilize several such sites in different geographic locations, so that even if, God forbid, the entire city gets wiped out by a nuclear weapon, they'll still be able to conduct business as usual. That's how much corporations value their immortality.

Sadly, though, there are no such protections in place for all of the precious information within our heads. All of the knowledge that we painstakingly acquire throughout our lives, all of the memories that we collect, all of the wisdom that accumulates with age; all of it is irrevocably lost once our brains get discarded. It seems a little remiss of us to say the least. If a technician were to start dismantling a computer without first taking a backup of the data, he'd probably get fired. Why don't we insist that surgeons follow the same procedure before operating on human beings?

No doubt the surgeons would probably consider that to be onerous impediment, since we currently don't have the means to create a backup of a person's brain. So what would it take to remedy that? The first thing we'd need to do is to figure out what needs to get backed up. This may be an overly simplistic way of looking at things, but in my mind there are essentially two data classes within the brain. The first involves action potentials, the electrical conversations that neurons use to talk to one another. When scientists use scanning devices to peer into a living brain, most of the time they are looking to detect action potentials, either directly in the case of electrodes, or indirectly via the hemodynamic response that's detectable in fMRI scans.

Action potentials are the currency of the brain, which means that as you're thinking about the words you are reading right now, those thoughts are represented in your brain in the form of action potentials. The ideas that reach your awareness are held in working memory as patterns of firing neurons. That doesn't mean that every neuronal firing in your head is directly involved in consciousness, but it is fair to say that action potentials are necessary for consciousness to occur. If your neurons are not firing, then it's not going to feel like anything to be you.

It's worth pointing out that when neuroscientists talk about neurons firing, they are usually concerned with significant firing events; in other words, events in which the firing rate is above a certain threshold.

There's always a certain amount of background firing going on, just to keep the neurons alive and healthy. If there's no firing activity at all, then you're probably dead.[27]

Most of the information that you keep in your head, however, is not stored in the form of action potentials per se. Action potentials are of course used in the storage and retrieval processes, but the data that holds our memories is actually stored in the dendritic trees, in the patterns formed by the synapses, or more specifically in the patterns of potentiated, or strengthened synapses.

It goes without saying that just because you have something in your head doesn't mean that it's on your mind. In addition to playing host to your conscious deliberations, your brain also contains all of the knowledge, experience and personality that constitute your identity. All of this content is stored away for safe keeping in the synaptic patterns, and it remains there until it's needed. When we remember something, it means we are pulling some memories out of storage, or more precisely, it means that the knowledge is now represented as a pattern of firing neurons, so at that point we're back to dealing with action potentials.

Despite the fact that we have machines that can take pictures of the inside of a brain, we do not yet possess the skills to capture all of the synaptic data. All we can get is a fairly crude representation of the brain activity that's happening at that particular moment. Even if we could somehow see all of the memory patterns, we'd have a hard time figuring out what to do with the data. Ideally, what we would like to do is to structure this data in the context of how the brain uses it, but this is difficult to do without really knowing the granular detail of how the brain is organized, and how it actually works.

How might we go about actually capturing the data? You're thinking nanobots, right? As you might have guessed by now, I'm not a big believer in nanotech heaven. Truth be told, I'm not a big believer, period, but it would be foolish, I think, to deny that it will ever be possible to build brain scanners with the necessary resolution, since we inevitably find ways to solve problems that are purely a matter of technology. I'm also not sure that it's a worthwhile exercise to try to guess the exact nature of future technologies, especially ones that are more than a couple of decades away. I remember how impressed I was when I first learned about the concept of using nuclear magnetic resonance for imaging applications, and I'll admit that it might have taken me quite a while to come up with that idea by myself.

Apart from our future goal of being able to extract the information content of our brains and make backups of our minds, there are, for sure, other good reasons for wanting to understand the circuitry of the brain. As medical science tries to grapple with more pressing neurological concerns, it certainly wouldn't hurt to have a better grasp on how the brain is wired together. We've begun to realize that we'll never really understand the normal functioning of the brain unless we make a determined effort to unravel the connectome – that mess of neural connectivity that somehow enables us to exist. As it happens, that's exactly the reasoning behind The Human Connectome Project, a government funded effort launched in 2009 to start tracing out those neural patterns in an attempt to get a better handle on the underlying framework of the brain.[28]

The project uses a new type of imaging machine, called a diffusion scanner, because it's able to detect the diffusion of water in the neural threads. They've now started calling it the Connectome Scanner and it works just like an MRI machine, except that the images it produces are about ten times sharper.[29] Although the task of mapping the connectome is still only in the early stages, I think we can be optimistic that in the next few decades, with incremental improvements in scanning technology, we'll start to see the emergence of a fairly detailed neural roadmap.

In addition to the new scanners, neuroscientists will continue to use other techniques to trace the neural pathways in the brains of animals, and to better understand the mechanisms that support neuroplasticity. Eventually, through the confluence of all this research, we should be able to build a decent working model of the brain that can fully explain the causal links between neuronal activity and cognitive functions such as consciousness and memory.

Assuming then that we were able to take a backup of the mind, we'd still have the problem of how to decipher the encoded memories. Neuropsychologist Steven Rose doesn't think this will ever be possible, because such information can only be interpreted in the context of the brain in which it is stored, and every brain is unique. Much of that uniqueness comes from the distinctive set of memories and experiences that have affected you during your life, which is why you and I will never have exactly the same perspective on things. That means, says Rose, we'll probably never be able to build a generic mind-reading machine like the cerebroscopes from science fiction.[30]

If Rose is correct, then what's the point in creating a backup? Even if I had a backup of Granddad Waine's brain, it's not as though I'd be able to play back his memories of what it was like growing up in the 1930s. We could digitize and store the content of a human brain, but without the brain itself there'd be no way for us to make sense of the data. So if that truck hits you, and your brain is turned to mush, what good is the backup if we don't have a way to restore the information?

What Rose is saying, it seems, is that the memory patterns from a human mind can only be meaningfully interpreted if they are invoked within the framework of a working brain – specifically, the brain that created the patterns in the first place. The rather obvious if not straightforward answer to this dilemma would be to construct a replacement brain. As long as we were able to configure the replacement using the data from the backup, we should be able to get things back the way they were before the mishap with the truck.

The new brain would most likely be configured during the fabrication process itself, so that we would end up with a replica of the original brain. The action potentials in the replica would then re-animate the data patterns from the backup, and as long as the new brain had the same causal powers as the original, would recreate that person's consciousness thus allowing the restored memories to be subjectively recalled.

I can hear my mom's voice saying that all of this sounds very far-fetched, which indeed it does. But if we're ready to accept that there will come a time when we have a good working model that describes how the brain produces the conscious mind, and if we're also able to build an advanced connectome scanner that can extract all of the information that's pertinent to the model, then the problem of how to create a brain becomes an engineering challenge that is likely to be met.

It's also worth noting that unless your name is Roger Penrose, we can probably agree that we won't have to worry about the quantum states of the atoms in our brains. More than likely, we'll be able to get all the data we need without going any deeper than ion channels and neurotransmitters, which means that our connectome scanner won't be nearly as complicated as the transporter device used in *Star Trek*.

Interestingly, on the question of whether it might be possible to create a brain, most of the top scientific and philosophical minds tell us that the answer is almost certainly yes, but that it won't happen anytime soon. The same folks also seem to agree that while it will not be easy to unravel the brain's mysteries, the problems are not intractable, and so they will eventually be solved. If the brain was fundamentally impossible to understand, then I doubt we would have quite so many gifted people working diligently on the case. The brightest of them would surely have given up by now. Instead, it's clear that progress is being made, despite the fact that scientists are restrained from disassembling a living brain to examine it.

In light of these conclusions, we are left with a very profound thought. What we've been discussing here are the essential ingredients in being able to cheat death, something that has never been possible in all of human history, despite many assertions to the contrary. If we can transfer the mind of a human being into a newly constructed brain, then we have found a way to place that person's identity beyond the reach of death. Certainly there are other details we'd need to take care of, since a disembodied brain is hardly what you'd call a victory over death, but once we can start replacing brains then the biggest hurdle is out of the way. From an engineering perspective, the brain is the most difficult part of the puzzle; everything else should be relatively easy.

But before we get too excited, there is another implication to what's being said here, one that is equally obvious, and just as profound. Although we can acknowledge that it is theoretically possible to escape death and to extend human lives indefinitely, it will not be an option within our lifetimes. So when Tyson asks the question "Can we live forever?" sadly the answer is no, at least as it applies to himself and me, and even Woody Allen for that matter. How unfortunate we are to be among the last few generations to live in the age of death.

At the risk of being labeled crackpots, let's imagine how things might look, say three hundred years from now, after we've perfected our connectome scanners and artificial brains. It's pointless to speculate about the materials that the brains might be built from, or whether they will be organic or completely synthetic, but one way or another they won't be much use unless they can interface with the outside world.

As you'd expect, the brain that I am using right now is connected securely to the rest of my body, and my various states of consciousness are being driven in part by the signals that I am receiving from various sources, such as the sensory input from my eyes and ears. I can feel my fingers on the keyboard, and as I type these words I can hear the clicking of the keys as I watch the characters appear on the screen in front of me. The fact that my butt has gone to sleep tells me that I've been sitting in one spot for too long, and I am also thinking that I could use a cup of hot tea right about now.

All of this information is being conveyed to my brain via chemical and electrical signals, and without these signals I'd be completely isolated from reality. There are also signals traveling in the outbound direction, most notably via the motor neurons, that allow me to move about and interact with the rest of you.

Likewise, the artificial brains of the future will need to be installed into bodies of some kind, so that they can get on with their lives. Maybe in three hundred years it'll even be possible to get an android body, like the one that was promised to *Star Trek*'s Lieutenant Uhura, except that this time you'll have an android brain to match. If you decide to move your arm, for example, your new brain would transmit the appropriate signals to a small computer which would then direct the motors in your arm to perform the intended action.

Believe it or not, this is exactly the kind of technology that's being developed right now in Brazil. Scientist Miguel Nicolelis is working on a brain interface that will allow quadriplegic patients to move about by themselves, using bionic leg braces that can be worn like an exoskeleton. The exoskeleton will be controlled directly by the brain, and will be fitted with sensors so that with every step the patients will be able to feel the ground beneath their feet.[31][32]

Considering that we've already started on the problem of getting signals to and from the brain, I don't believe that brain interface technology will be hugely challenging for twenty fourth century engineers, especially if they already have the means to build actual synthetic brains. We can assume I think, that these brain interfaces will be well defined, and that the messaging streams could be coordinated by computers of some sort, in which case we might not even need physical bodies at all. Instead of sending messages to and from a mechanical contraption, why not allow the computers to orchestrate the messaging in a way that enables the brain to interact with a virtual world instead of the real one?

What I'm proposing here is similar to the sort of virtual existence predicted by others, though with a few differences. First of all, unlike the AI enthusiasts, I'm not suggesting that our minds will be emulated via software, although in the long run this may turn out to be a somewhat irrelevant distinction. A more significant difference is that whereas singularitarians prophesy that nanobots and intelligent computers will make all of this happen by 2040, I'm thinking it will take a lot longer.

While virtual reality may not be everyone's cup of tea, my guess is that it will become the default choice for a number of reasons. Having to outfit every individual with a real body would hugely expensive, not just in terms of the resources needed to build and maintain the bodies themselves, but because we'd all still be moving about in the real world, using up real estate and raw materials to create all of the physical commodities that we desire.

If instead we can encourage people to pack up and move all of their worldly possessions into virtual space, then it greatly reduces our environmental footprint, and it also greatly simplifies things from an engineering standpoint. Android bodies need to be designed to work in the real world, where the laws of nature can be very unforgiving. In a virtual setting, however, engineers will have the convenience of being able to change these laws if necessary, which gives them more control over the rules of the game.

The concept of virtual reality is not exactly new. When my kids were younger, they liked to frequent a place called Poptropica, where they could explore interesting islands and interact with other kids. They could also use Poptropica credits to purchase clothes and other items to accessorize themselves, which I believe was the main attraction for my daughter. These days, they seem to spend an inordinate amount of time wandering through the pixelated landscape of Minecraft, where they can build all kinds of wondrous things, including elaborate castle dwellings to house all of their pixelated belongings. Cyberspace is not just for kids either, as a lot of adults also seem to be taking it seriously. A work colleague once told me that he was leaving early to help set up an art exhibition in a shopping mall. It sounded interesting so I thought I might go along, but when I asked for the address he gave me a URL.

I can't say that I've ever actually visited one of these places, and while they sound like fun, I'm not sure that I'd like the idea of actually taking up permanent residence on one of Poptropica's islands. Then again, if it were offered as an alternative to complete annihilation, maybe I'd think about it.

While today's virtual reality might not sound like anybody's idea of paradise, how might things look from a twenty fourth century perspective? One thing we seem to be getting better at is faking things. Our movies and video games are starting to look incredibly realistic, due in large part to advances in computer animation techniques. We're also seeing exciting new developments in immersive 3-D capability, and given the rate at which these technologies are improving, it seems like a safe bet that the virtual world of tomorrow will eventually become indistinguishable from reality.

Suppose you could live in a virtual world and go snowboarding in the Sierras and then drive to Yosemite to spend a weekend with your family. What if being able to walk on a virtual beach was identical in every respect to the real experience? You'd be able to feel the waves splashing over your feet, the warmth of the Sun on your face and the refreshing breeze in your virtual hair. Virtual reality is sounding a bit more palatable, don't you think?

We wouldn't have to worry about huge wildfires raging out of control, or buildings getting leveled by earthquakes; no more agonizing about asbestos exposure, freak accidents, or lives being curtailed by disease. There'd be no need for people to suffer from chronic conditions such as back pain, or acid reflux. Quality of life would be immeasurably better. Once you get used to it, the virtual world idea begins to grow on you.

It should also open up new possibilities for space exploration along the lines that Tyson suggested earlier. We could explore the galaxy without having to put lives at risk. And we'd no longer have to worry about getting hit by that truck.

From an environmentalist's standpoint, it solves a bunch of problems right off the bat. Arguably the biggest issue facing the planet has been the steadily increasing human population, and the huge burden that this places on our global ecosystems. Many fear that by eliminating death, we'll be just exacerbating the problem by accelerating population growth. Ironically though, it may turn out that the best way to save the planet may be to save ourselves.

If a significant portion of humanity were to migrate into a virtual community, then over time this would certainly translate into an overall decrease in the quantity of resources extracted from the environment. We'd still need energy to support the banks of artificial brains, and the servers hosting the virtual worlds, but when you consider how much real estate and natural treasure that's currently consumed just to keep humans happy, this would definitely be a good deal for nature. We could still support a large population but without crowding up the place, and we might even envision a scenario in which wildlife habitats are restored, and much of the surface is given back to the non-human species with whom we are supposed to be sharing the planet.

It's not clear whether our numbers would keep growing once we start making the transition to cyberspace. Maybe the birth rate would plunge dramatically and the population would level off, similar to what we see happening in parts of Europe. Fortunately though, we'll no longer have to deal with the problems of aging. I know you're probably wondering how people can have babies in virtual reality, but I think I'll leave that one to your imagination.

Now at this point, if you're buying any of this, you may be thinking that the future I'm describing is one in which all semblance of human decency has been thrown out the window, and that everything has gone to hell in a hand basket. Here we are talking about creating babies in cyberspace, for heaven's sake. How can that be good?

Strange as it might sound though, it seems to me that in spite of, or perhaps even because of these remarkable changes in society, people will find themselves becoming even more conservative, and will make every effort to hold onto their traditional values. When the ancient Egyptians developed the concept of an afterlife, they might have dreamed up an existence that was completely different from reality; one that was filled with all manner of exoticisms. Instead, their version of heaven was very similar to the life they already enjoyed; a magical place that bore a close resemblance to the Nile valley, where people continued to farm the land just as they had done in the past, except that, conveniently enough, they no longer had to put up with pests, and their harvests were more bountiful. That was, after all, the whole point of having an afterlife – to hold on to the things that were important.

Likewise, if our descendants ever decide to migrate to a virtual world, it will not be because they are fed up with life. It will be for precisely the opposite reason, so that they can preserve all of the things that they love about the real world.

Many futurists have speculated that virtual reality will become a sort of Wild West, where anything goes. Ray Kurzweil tells us that we can all look forward to multiple sex partners, and that we can engage in all manner of debauchery without having to deal with the consequences of STDs.[33] Since the computers will be able to conjure up anything we desire, we'll essentially be able to do whatever we want, so the implication is that we will adopt a philosophy of moral ambivalence and embark on an endless rampage of self-indulgence. Even the laws of physics will be negotiable, so we can all sprout wings and fly around like birds, if we so desire. Perhaps we will do all of those things, but that hardly seems like an adequate summation of where society is headed.

There is a place for debauchery, I'll not deny it. As Kurzweil is aware, we all have a craving for sex and entertainment, which is probably why he dwells on it in his book. It's a fair observation, I think, that among those who would champion the idea of living in cyberspace, there does seem to be a disproportionate emphasis on the idiosyncrasies of virtual life.

The truth is, though, we've been living with our cravings for quite a while now – for all of human history, in fact – and yet our appetite for novelty hasn't become the dominant force in society. We have Las Vegas, of course, which is a great place to visit when you're looking for distraction, but after a while most of us are happy to get back home to some sense of normalcy.

The geeks would have us believe that as soon as technology removes the constraints on what's possible, we're all just itching to throw away the rules and reinvent ourselves. What's missing here is the realization that we could have thrown away the rules a long time ago, but we chose not to, which is why we don't all live in Vegas.

Rules may seem like a bad idea at first, but without them we would lose our identity, and civilization would cease to exist. Our cultural and societal values are based upon rituals and traditions, which in turn are dependent upon a generally accepted set of rules and principles. These rules provide a framework for human behavior that incorporates recognized moral standards and codes of conduct, as well as institutionalized laws and customs. In other words, we expect everything to work in a certain way, and we like it when it does.

Granted our societal values can change and evolve but this usually happens as a result of social interaction and consensus, and not purely as a result of technological progress. When it comes to how we feel about ourselves, and the rules that we consider to be important, we've never really been constrained by technology, or by the limitations of the natural world for that matter. In a sense, because of our unique human ability to reimagine ourselves, we've have always lived a virtual reality of sorts; a world of symbols, meaning and tradition; of gods and superstition, and everlasting life.

This way of living then gets passed down to subsequent generations via the rules and rituals of our culture. It would be foolish, I believe, to underestimate the importance of this heritage, because if that happens, if in the future we get so caught up in the excitement of technology that our lives become consumed with frivolities, then there is a danger that we could lose ourselves as a result. How tragic it would be if the path to salvation were to lead to our ultimate downfall.

I don't think that will happen though, and I'll explain why. Let's take soccer for example. Granddad Waine was very fond of soccer or football as he would have called it, and I've heard that he was quite a talented player in his youth. A couple of months before I was born he went to see the final of the World Cup, and I'm told he was overjoyed when England took care of Germany. In his later years, he used to visit my parents once a week to watch the Saturday games on their color TV.

What you notice about soccer, or any other sport for that matter, is that most people generally comply with the rules. There are no technology constraints or laws of physics that would prevent someone from picking up the soccer ball and running with it, but we decide not to do that because we know that it would just mess up the game.

If Granddad were alive again, we could sit down together to watch an Everton match, and even though he hasn't been around for almost forty years, he'd still be able to recognize and appreciate what was going on.

Nothing much has changed, apart from the fact that we'd be watching a wide format broadcast in crisp high definition. I think he'd be impressed by that.

Generally speaking, humans are all about social interaction, and so in a sense, all of reality, virtual or otherwise, is just a backdrop for the human conversation. Within this backdrop, we are sure to preserve the parts of our heritage that we are truly passionate about. If you're a baseball fan, then you've got to believe that we will keep the World Series, because otherwise life wouldn't be worth living. As we know, there are other people who are passionate about the natural world, and about politics and civility, justice and the preservation of human rights, so no doubt all of these interests will become part of our enduring legacy as well. And so, for the most part, our descendants may very well choose to continue playing by the same rules, whether they are motivated by passion or just by simple nostalgia.

The transition to a virtual existence is obviously not something that could happen overnight. It will be a long, gradual process of adjustment. Society is already doing its best to cope with the fact that so much of our culture is moving over into cyberspace, but it will still take a long time, I think, to adjust to the idea of people literally living there too.

In many ways, we are already laying the groundwork for this transition. A lot of business activity is now conducted exclusively in the virtual domain. My wife and I do a lot of our shopping there, although there are some things that I prefer to buy in a store – a non-virtual one, that is. When selecting an item of clothing, for example, it's nice to be able to try it on for size. I once tried purchasing a pair of shoes online, but I had to return them three times before I got a pair that would fit me.

A co-worker tells me that at some stores you'll soon be able to step into a measuring booth, something like the body scanners that are used by airport security. These machines will do a complete scan of your body and then feed all of your measurements into an online database. The idea is that when you log on to the retailer's website you can pick out the clothes you want, and then get your avatar to try them on before you buy them.

It sounds like a fine idea to me, although as we continue down this path it becomes a problem if we end up spending too much time sitting in front of the computer. I'm glad that my kids are able to navigate their way around Poptropica and Fantage, but I also need them to be outside, getting plenty of exercise, fresh air, and vitamin D from the Sun. As long as we still have physical bodies to take care of, it's unhealthy to get too preoccupied with the virtual world.

Perhaps our descendants won't have to worry about exercise anymore, once they relinquish their corporeal existence and cross over to the new frontier that awaits them. Although we cannot make the journey ourselves, we can watch this virtual realm begin to take shape, as though peering through a window, or in the case of Microsoft, many such windows.

The good thing is that even though we can't physically get to the new frontier, we can keep all our stuff there. Many of us now prefer to get everything in electronic format, including invoices, receipts, airline tickets, books, photographs, movies and music. Not only is it better for the environment, but it also means that everything is secure and accessible, and you don't have to lug a big briefcase around.

Given the frantic pace at which we're digitizing everything, there should be plenty of content available in virtual land by the time people start living there for real. We've gotten used to the idea that the best way to preserve something is to digitize it. Anything considered important from a scientific or artistic perspective inevitably gets reduced to zeroes and ones and shipped off to cyberspace.

Eventually we'll digitize our cities as well, perhaps starting with the great cities of civilization, such as Rome and Paris. Regardless of whether people end up living in simulated bodies, there's an obvious reason to want to create virtual replicas of culturally significant places. Physical artifacts slowly deteriorate with age if they are not properly maintained, and they can also suffer catastrophic damage due to earthquakes or deliberate vandalism. That's what happened in 2012 when the Bologna area of Italy was hit by two major earthquakes over a period of ten days. Twenty five lives were tragically lost, and a number of historic buildings were damaged or destroyed.[34] [35]Even if a building manages to survive for centuries, it can still be reduced to rubble in a matter of seconds.

It's true that the media types used to store digital information will also degrade over time, but it's also true that the information can be readily refreshed, transferred to a different medium, or even reconstructed using a new encoding scheme. If and when our kids discover better ways to represent information in the future, they should be able to convert the digital data into whatever format they come up with, even if it's analog.

Thus, if all goes well, our descendants will preserve the rich cultural heritage of humanity in a virtual heaven, where no one ever dies. Perhaps this can be regarded as an appropriate and worthy development, especially since the world of human culture has always been one of our imagining, a symbolic world that has only ever existed in the virtual space created by our minds, and in that sense, detached from the base realities of nature. With regard to the things of the world, it is the symbolic meaning that concerns us, and not so much whether they are real or virtual.

As we've come to understand the story of humanity, this propensity for symbolic thought is what characterizes us as human beings. It is what allows us to act in defiance of nature, and distinguishes us from the other beings that share the planet with us. We have distanced ourselves from their reality, and we no longer subscribe to all of Nature's rules. As John Muir disapprovingly noted, humans have always regarded the natural world as simply a stage for human society to play itself out. Our story has always been painted on a canvas that's much broader in our imagination and intellect than the backdrop of nature, with its cold, uncaring veneer of death and regeneration. Soon, it seems, we are destined to become completely immersed in the world we have imagined for ourselves.

So let's just take a moment to summarize how we came to this conclusion:

For this scenario to play out, we really just need four things to happen. First, we need artificial brains. The second thing we need is a way to copy our minds into these artificial brains. Thirdly, we need a rich virtual reality environment, and then the fourth step would be to directly interface the artificial brains to the rich environment.

The first challenge is probably the most difficult, and yet nobody can come up with a good reason for saying that it'll never be done. Even the philosophers are nodding their heads about it only being a matter of time. If we're looking ahead three hundred years, artificial brains seem like a reasonable goal, provided we stay on the carousel of progress and don't destroy ourselves anytime soon.

The second challenge goes along with the first to some extent, the assumption being that if you can build an artificial brain, then you've probably been able to build an advanced connectome scanner that can extract the salient data needed to configure the artificial brain so that it reanimates the mind. It's a lot of speculation, I know, and as Donald Rumsfeld would say, "There are also unknown unknowns – the ones we don't know we don't know."[36] But again, at this point there's no reason to see this as anything more than a difficult engineering problem, and given the possibility that it could allow us to extend human lives indefinitely, it should be one that engineers will be properly motivated to solve.

The third challenge is being addressed as we speak, since virtual reality is already reality, and it's a technology space that's certain to see continuing progress even in the near term, let alone three hundred years. The fourth task is to hook up the brains to the simulated environment and if we somehow managed to meet the first two challenges, then this last step should be up relatively easy, especially when you consider that scientists are already actively working on the goal of building interfaces to the brain.

It's a lot of change to get used to, that's for sure, but this is what happens when we embrace the philosophy of progress, and the promise of a better future through technology. Perhaps we shouldn't be surprised when we get what we ask for. With the way things are headed, frankly it would be harder to imagine a future in which people choose to keep everything the same, avoiding such abominations as artificial brains and fake realities.

Whenever we have to accept some unwanted consequence, we say that it's just the way things are. There was a time, however, when "the ways things were" meant that people died from appendicitis, but now they don't. Perhaps sometime soon, people will no longer die from lung cancer, and so "the way things are" will have changed again. We've all subscribed to this idea of change, and the changes that occur are due to the choices that people make. It seems to me a no-brainer that people will continue to choose life rather than death.

Still though, I expect that for some people, simulated bodies and artificial brains will be too much to take, especially for those apologists for death who insist that it is our mortality that defines us. They may reject the idea of life prolongation, but unless they can convince everyone else to go along with their plan, theirs is likely to be a short lived philosophy.

Writer and philosopher Stephen Cave has an interesting take on all of this. He claims that even if it is someday possible to have your brain copied, this wouldn't help you to avoid death because the new brain is, after all, just a copy. It wouldn't be the real you.[37] So that got me wondering about whether there is such a thing as the real me. From what I've learned, the essential ingredients of the real me, namely consciousness and memory, are due to the activity that's happening inside my brain, at least according to another philosopher, John Searle. Searle also tells me that my sense of identity is obviously derived from the memories that I have access to, and the fact that these memories remain somewhat consistent from one moment to the next.[38]

I realize, though, that when I wake up in the morning, my mind is never in the same state that it was in just before I went to bed. I'm not thinking the same thoughts, I'm not aware of the same percepts, and I'm not activating the same memories. So am I still the real me, or do I keep changing into a different person? Strange though it may seem, I think it's the latter. I am a different me than the one that existed yesterday, or last week, or last month. The further back you go, the easier it becomes to notice the differences. All of those other versions of me are now dead and gone, but that doesn't bother me, because I am here in their stead.

I'm constantly adding to my storehouse of memories, and it's these memories that provide the thread of continuity that gives me my sense of identity. Notice how I just repeated myself? That's because the memories are not perfectly preserved, and I tend to forget things every now and then. Nevertheless, I find that the process works well enough to meet my requirements. If you were to inquire as to whether you were talking to the real me, I would probably always answer in the affirmative, and I expect the same would be true if you could somehow make a copy of me. The copy would also believe himself to be the real me, and he'd be correct. Although, to be fair, he'd have to admit that I was here first.

If you did ever create a copy of me, the two versions would then begin to diverge, so that after a while we'd no longer be exactly the same. Maybe this is what Cave is worried about, but since most of us would prefer not to have copies of ourselves walking around, I can't imagine that this is something that people will choose to do in the future. I suspect our home owners association would object to the idea. On the other hand, it might be nice if I could take a backup of my mind each day; in case it happens to be the day I get hit by the truck. If I end up dead, then you could install the backup into a new brain and bring me back online.

If this brand of immortality isn't good enough for Cave, then I think he may be setting the bar a little too high. In my book, if you can create a thinking, remembering version of me, with the same subjective sense of actually being me, then I'd say you've earned your pay for the week.

I guess this means we've reached the end of our quest, since we're finally able to answer the questions that we posed at the outset. On the matter of whether it's possible to defeat death, the answer would seem to be yes, although not just yet. And as much as I'd hoped that there might be a way to bring Granddad Waine back to life again, there doesn't seem to be a way to do this. Sadly, it looks like the dead will have to remain dead.

What about the question that Gilgamesh asked at the beginning of his journey: "Must I die too?"[39]

Unfortunately for us, the answer to that question is yes. There is no escape for the likes of you and me. I know this must come as a bit of a disappointment, and in that sense I suppose the same could be said for the Gilgamesh epic itself. You might have expected that the story would have some climactic ending; a battle between good and evil, or some amazing heroic deed, or perhaps even the death of the great Gilgamesh himself. In the end, however, nothing happens. When Gilgamesh finally realizes that his search for immortality is getting him nowhere, he returns home. That's it. The story ends where it began, just outside the walls of the city, where the king makes his closing remarks.

> This is the wall of Uruk, which no city on earth can equal. See how its ramparts gleam like copper in the Sun. Climb the stone staircase, more ancient than the mind can imagine, approach the Eanna Temple, sacred to Ishtar, a temple that no king has equaled in size or beauty, walk on the wall of Uruk, follow its

> course around the city, inspect its mighty foundations, examine its brickwork, how masterfully it is built, observe the land it encloses: the palm trees, the gardens, the orchards, the glorious palaces and temples, the shops and marketplaces, the houses, the public squares.[40]

There's no more talk of immortality or great adventure. It's time to return home, and to get on with life; or as the poet T.S. Eliot says, to "arrive where we started, and know the place for the first time."[41]

Cave concludes that, like Gilgamesh, we should forget about immortality. We need to snap out of it, he says. We're going to die, and that's just the way things are. But then he wonders how we're all going to cope. How are we to console ourselves given the news that death is inevitable?[42]

Maybe, says Cave, we'll feel better if we imagine that our lives are like books. A book has a beginning, a middle and an end, and it's the end that makes it all worthwhile.[43] It just wouldn't make sense for things to continue on indefinitely. I suppose the argument is that we can pick up a book and read it over again, just as we can review the episodes in a person's life.

The problem with that analogy is that life isn't like a book, at least not from the point of view of the person who's living it. Physicists will tell you that the passage of time is an illusion, and that past, present and future are all laid out before us in the description of the universe, like the pages of a book. One such physicist, Paul Davies, says that this should make us feel better about dying.[44] Yet it seems to me that a human being can only exist by moving forward through time. We are trapped in the present. The subjective experience of being me can only happen while I'm moving into the future, converting the present into the past through the creation of memories. As Davies admits, "the formation of memory is a unidirectional process," which means that subjective experience is intimately linked to our perception of what we believe to be the flow of time.[45] That being the case, it's hard to understand why we should feel better if we imagine that time doesn't really flow at all.

As you read this book, the conversation is all one way; all of the conscious activity is in your head, not mine. Cave implies that when we die, our lives will suddenly be endowed with meaning, but from my point of view my life only has meaning while I'm living it. When I reach the end of my life, it will have no further meaning as far as I'm concerned, for the simple reason that I won't be able to concern myself with anything. Searle explains why.

> Consciousness is the condition that makes it possible for anything at all to matter to anybody. Only to conscious agents can there ever be a question of anything mattering or having any importance at all.[46]

If a person is like a book, then a human life is nothing more than just two dates with a list of occurrences in between. For Cave, there may be some meaning in there somewhere, but how could any legacy ever be worth the loss of life itself, and that wonderful feeling of being alive?

As you may have surmised, I don't quite agree with Cave on that point. We're not defined by death; we are, in fact, defined by an attitude toward life that says we're not going to accept things the way they are. Death is a wasteful process that has no regard for the preciousness of human experience, or the value of the information inside our heads. Personally speaking, the chronology of things that happened in my life is not as important to me as the fact that I am able to enjoy life right now, in the present, spending time with my family. To derive meaning from life, we must be alive.

Yes, we're going to die, and despite Cave's fear that we won't be able to handle this revelation, I don't think he needs to worry. People have had to live with death for a long time and somehow or other they've been able to deal with it. Many of us start out with the same concerns as Gilgamesh, and in the end, like the great king himself, we decide that it's better not to dwell on the subject. My dad tells me that he used to worry about death, but now, he says "I don't give a fiddler's," which is an Irishman's way of saying that it no longer concerns him.

Then there are the folks to whom the phrase "ignorance is bliss" might be applicable. Psychologist Jesse Bering discovered that although most of us know that we're going to die, we don't necessarily have a good feel for what this actually means. Bering describes a study that he conducted on a group of students, many of whom were self-professed "extinctivists"; a term he uses to describe people who hold that heaven doesn't exist, and that death is simply the end of life. He gave them a questionnaire that asked a series of questions about the supposed mental state of a man who had just been killed in an automobile accident. When he asked if the dead guy, Richard, was aware that he was dead, Bering was surprised to find that a lot of the participants answered in the affirmative. One of the extinctivists even got impatient with Bering, and declared "Of course Richard knows he is dead, because there's no afterlife and Richard sees that now."[47]

As Bering observed, his subjects seem to be having difficulty understanding that dead people don't have conscious states, so they aren't aware of anything. Apparently it's not easy for us to grasp the reality that dead people are really gone. "Back when you were still in diapers, you learned that people didn't cease to exist simply because you couldn't see them," he explains.

For example, he can imagine his friend Ginger who lives in New Orleans, and in his "mind's eye" he can see her "walking her poodle or playfully bickering with her husband." Our brains get so good at this, says Bering, that when people die, it's hard for us to get used to the fact that they no longer exist at all. "We can't simply switch off our person-permanence thinking just because someone has died."[48]

I suppose it's not a bad thing that we all have our own peculiar ways of thinking about death. One way or another, we all have to find a way to live with the fact that we're going to die, and most of the time we manage to shove it to the back of our minds, but as Koch admits, every now and then we get a shock when the reality sinks in.

> I suddenly woke up in the middle of the night knowing I was going to die: not right there and then, but someday.... Just the gut realization that my life was going to end, sooner or later.[49]

Somehow it doesn't seem right, does it? Having made it to the twenty first century, are we really saying that we don't have a better answer than the one Gilgamesh came up with forty five centuries ago? Like Benjamin Franklin, are we still just as certain about "death and taxes"?[50]

Well, there is another path available to us, one that wasn't an option when those two gentlemen were around. It's called cryonics, and the idea is that you can have yourself frozen in the hope that some resourceful and altruistic engineer of the future might come up with a way to make you alive again.

The Alcor Life Extension Foundation offers two pricing plans: You can have your head frozen for $80k, or you can pay $200k and get them to freeze your whole body. If you're thinking about signing up, I would probably go with the cheaper plan, since the information you want to preserve is mostly in your head. Some will argue that to be properly resurrected you need to save more than just your head, because the brain receives a lot of important communiqués from the rest of the body. That's true; however these signals tend to drive the brain to different states, so all the necessary information should theoretically be preserved in the brain.

Alcor refers to the frozen heads as patients, since, in theory, they are only temporarily dead. The way the process works is that they first take the head and inject in some antifreeze which is supposed to prevent the formation of ice crystals that might damage the brain. The antifreeze probably does some damage too, but maybe when compared to ice crystals it's the lesser evil. The head is then brought to a temperature of -196 degrees and stored in a vat of liquid nitrogen, where it remains indefinitely until some future geeks are ready and willing to revive the patient.

To perform the resuscitation, our clever descendants might use a device similar to the advanced connectome scanner we mentioned earlier, to scan the preserved brain and extract all the needed patterns representing the memories of the deceased. To perform this task, everything will need to have been well-preserved. In particular, the hope would be that all the neural wiring will still be intact, including all of the synaptic patterns. Presumably the future folks will need enough information to be able to distinguish between the potentiated and non-potentiated synapses as well. If there's been too much degradation and the necessary biochemical detail is not present, then the resurrection process won't work, no matter how sexy the technology might be.

In the happiest scenario, once they've extracted the information from the frozen mush, they will then be able to fabricate a new brain with exactly the same neural configuration and synaptic patterns as the old one, and then reanimate the dead person's mind. It's commonly assumed that the resurrection technique might involve restoring the actual defrosted brain itself, although personally I think that would be a little messy and probably unnecessary. I'm betting the scanning will be done non-invasively, which means there won't be any defrosting or dissecting required.

I'm reminded of a procedure carried out in 2011 by a team from the Lawrence Berkeley National Laboratory, which admittedly had nothing to do with scanning frozen brains to recover the minds of dead people. Rather in this case the challenge was to retrieve the long forgotten personality of a talking doll. In the late nineteenth century, the famous inventor Thomas Edison came up with the idea of placing a phonograph mechanism into a child's doll. He employed a young lady to shout the words of a nursery rhyme while he recorded her voice onto a small metal ring which was built into the mechanism. The doll could then be made to speak by turning a crank that protruded from its back. The idea was a flop, but somehow one of the little rings managed to survive, and had been lying around in a museum for years. Nobody had any clue as to what kind of recording it might contain, if any, as the ring was badly damaged and completely unplayable.

Using a new 3D optical scanning technique, the LBL folks managed to extract the voice data from the ring, and they then fed this data into a computer which cleaned up the distortions and played back the original sound. So, without touching the doll or the ring itself, they were able to reanimate its mind, so to speak. The analogy is a bit of a stretch, but you get the idea.

Restoring a human mind would be a much greater challenge, of course, but neuroscientist Sebastian Seung thinks that, in theory, it might be possible. It all depends upon whether a person's connectome can properly preserved within a frozen brain, but right now, Seung admits, we don't know enough to be able to answer that question. Once we know how to identify the connectome, then we could do some tests to see how well it gets preserved. All he can say for now is that there's a "small but nonzero" probability that cryonics will work.[51] Given that so much of this is just guesswork, most scientists are not even willing to speculate on whether any sort of resurrection might ever be possible. I guess if you have bags of money, and you trust the folks at Alcor and you're willing to take a leap of faith, then what have you got to lose? Go for it!

Kübler-Ross was quite critical of the notion of cryonics, because she felt that resurrecting people would just add more bodies to what would eventually become a hopelessly overcrowded world.[52] But if the newly resurrected are willing to live in cyberspace and forgo actual bodies, then maybe that wouldn't be an issue after all.

Aside from the slim chance of longevity offered by the cryonics crapshoot, it's fair to say that the rest of us will share the same fate as our friend Gilgamesh, but it's also true that we are no longer quite so clueless about the nature of our existence. If the ancient king were alive today, he would surely admit that we have come a long way in our quest for answers.

Some years ago I was sitting on a visitor bus at the Johnson Space Center in Houston, listening to the automated tour guide as it espoused the virtues of space exploration. The narrator described how men and women would continue to explore the new frontiers, and that maybe someday we'd get some answers to the big questions; that we may even discover who we are, and where we came from. It occurred to me that we actually have a lot of those answers already. I'm not suggesting that there's nothing more to be discovered, but as Searle points out, "Long standing philosophical problems such as the nature of life have been, I think, pretty much decisively resolved."[53]

Among the things that remain mysterious, says Searle, are the origin of life and the functioning of the brain. Yet even on those topics, we have a reasonable grasp of the underlying principles and mechanisms at work. Gilgamesh wouldn't have had the foggiest notion of how the mind was produced, whereas now we're talking about synapses, neuroregulators and the like. Likewise, on the origin of life, we can at least boast about having the gist of the story. We can speak with certainty about the age of the planet, and about what the universe was like in its early stages. We can put together a comprehensive storyboard describing how the solar system came into existence, and how life developed and evolved over time from simpler beginnings. We can describe how genetic information gets passed from one generation to the next, and we even have an answer to Gilgamesh's big question – is there a way to beat death? We can say, with some degree of confidence that the answer is yes, or at least it will be for those denizens of the far future.

What will they think of us? I wonder. If our species can survive beyond this critical juncture, then you and I are participants in the final act in the brief prologue of civilization. We are the last to live in the Age of Death. To our immortal friends in the future, we can only ask that they remember us well. Perhaps they will. Perhaps they will bring our stories to life again, just as we enjoy reliving our past, and remembering with nostalgia the birth of our civilization in the Fertile Crescent, and the glorious days of the great king Gilgamesh.

> Twelve hundred million men are spread
> About this Earth, and I and You
> Wonder, when You and I are dead,
> What will those luckless millions do?[54]
>
> - Rudyard Kipling, "The Last Department"

BIBLIOGRAPHY

"Changing Your Mind." *Scientific American Frontiers*. Written, produced and directed by Graham Chedd. 55 min. The Chedd-Angier Production Company, Inc., airdate: 21 November 2000.
"Chemical Controls." *Scientific American*, vol. 302, no. 4, April, 2010, p. 30.
"Fourth Report of the New York State Factory Investigating Commission." *Documents of the Senate of the State of New York Legislature: One Hundred and Thirty-Eighth Session*, vol. 14, no. 43, part 5. Transmitted to the legislature, February 15, 1915. Albany, New York: J. B. Lyon Company, Printers, 1915.
"Fractals: Hunting the Hidden Dimension." *Nova*. Produced and directed by Michael Schwarz and Bill Jersey. 56 min. A Quest Productions and Kikim Media Production for Nova in association with The Catticus Foundation, 2008. DVD Video.
"Genetically Modified "Serial Killer" T Cells Obliterate Tumors in Patients with Chronic Lymphocytic Leukemia, Penn Researchers Report." News Release, 10 August 201. Philadelphia: University of Pennsylvania, 2011. Available from http://www.uphs.upenn.edu/news/News_Releases/2011/08/t-cells/
"Good Luck, Father Ted." *Father Ted*, season 1, episode 1. Directed by Declan Lowney. 25 min. Hat Trick Productions, 1995, DVD Video.
"I, Mudd," *Star Trek: The Original Series*, season 2, episode 8. Directed by Marc Daniels. 50 min. Desilu Productions, 1967, DVD Video.
"IBM Unveils Cognitive Computing Chips." News Release, 18 August 2011. Armonk, New York: IBM Media Relations, 2011.
"John Muir in the New World." *American Masters*. Directed by Catherine Tatge. 1 hr. 25 min. Produced by THIRTEEN for WNET, 2011. DVD Video.
"Mudd's Women." *Star Trek: The Original Series*, season 1, episode 6. Directed by Harvey Hart. 50 min., Desilu Productions, 1966, DVD Video.
"NIH Launches the Human Connectome Project to Unravel the Brain's Connections." *NIH News*, 15 July 2009. Available at http://www.nih.gov/news/health/jul2009/ninds-15.htm
"Relics." *Star Trek: The Next Generation*, season 6, episode 4. Directed by Alexander Singer. 46 min., Paramount Television, 1992, DVD Video.

"Second Chances." *Star Trek: The Next Generation*, season 6, episode 24. Directed by LeVar Burton. 46 min., Paramount Television, 1993, DVD Video.

"Slim Chance of Tuning in to ET TV." *New Scientist*, vol.195, p. 16; August 4, 2007.

"Soviet Reverses Position." *Astronomy*, vol. 5, no. 1, p. 56; January 1977.

"Tentacles of Doom." *Father Ted*, season 2, episode 3. Directed by Declan Lowney. 25 min. Hat Trick Productions. 1996, DVD Video.

"The Changeling," *Star Trek: The Original Series*, season 2, episode 3. Directed by Marc Daniels. 50 min. Desilu Productions, 1967, DVD Video.

"The Human Exploration of Space." Washington, D.C.: National Academy Press, 1997.

"The Measure of a Man." *Star Trek: The Next Generation*, season 2, episode 9. Directed by Robert Scheerer. 46 min., Paramount Television, 1989, DVD Video.

"The Neutral Zone." *Star Trek: The Next Generation*, season 1, episode 26. Directed by James L. Conway. 46 min., Paramount Television, 1988, DVD Video.

"There's a Great Big Beautiful Tomorrow." Original theme song for Walt Disney's "Carousel of Progress." Written by Robert B. Sherman and Richard M. Sherman. Published in 1964.

"Unsatisfactory Conditions Found in Syracuse Plants." *The Post-Standard*. Syracuse, New York, November 30, 1911.

"War." Written by Norman Whitfield and Barrett Strong. Released 10 June 1970. Performed by Edwin Starr.

Ackerman, Diane. *An Alchemy of the Mind: The Marvel and Mystery of the Brain*. New York: Scribner, 2004.

Alda, Alan. "Fat and Happy." *Scientific American Frontiers*. Written, produced and directed by John Angier, David Huntley and Andy Liebman. 55 min. The Chedd-Angier Production Company, Inc., airdate: 1 May 2001.

Anderson, Jesper L., Peter Schjerling and Bengt Saltin. "Muscle and the Elderly." *Scientific American*, vol. 283, no. 3, September 2000, p. 54.

Anderson, Per, et al. (eds). *The Hippocampus Book*. Oxford; New York: Oxford University Press, 2007.

Aries, Philippe. The Hour of Our Death. First Vintage Books Edition. Translated by Helen Weaver. New York: Random House, Inc., 1982.

Assmann, Jan, trans. *Death and Salvation in Ancient Egypt*. Translated from the German by David Lorton. Cornell University Press, 2005.

Attenborough, David. "Social Climbers." *Life of Mammals*, season 1, episode 9. Produced by Mike Salisbury and Huw Cordey. 50 min. BBC, 2003, DVD Video.

Babe. Directed by Chris Noonan. 1 hr. 34 min. Kennedy Miller Films, 1995, DVD Video.

Bailey, Joseph K, et al. "Fractal Geometry is Heritable in Trees." *Evolution*, vol. 58, no. 9 (2004): pp. 2100-2102.

Bartlett, Albert A. "Forgotten Fundamentals of the Energy Crisis." *American Journal of Physics*, vol. 46, no. 9, September 1978, pp. 876-888.

Bazell, Robert. "New Leukemia Treatment Exceeds 'Wildest Expectations.'" *msnbc.com* [online], 10 August 2011. Available at http://www.msnbc.msn.com/id/44090512/ns/health-cancer/t/new-leukemia-treatment-exceeds-wildest-expectations/

Beardsley, Tim. "Gene Therapy Setback." *Scientific American*, vol. 282, no. 2, February 2000, pp. 36-37.

Becker, Ernest. *The Denial of Death*. New York: Simon & Schuster, 1997.

Bell, Robin E. "The Unquiet Ice." *Scientific American*, vol. 298, no. 2, February 2008, pp. 60-67.

Berger, Theodore, et al. "A Cortical Neural Prosthesis for Restoring and Enhancing Memory." *Journal of Neural Engineering* 8 (2011) 046017 (11pp)

Berger, Theodore. Panel discussion during the Cognitive Computing conference, hosted by the IBM Almaden Institute, 10 May, 2006. Video download available at http://www.almaden.ibm.com/institute/2006/agenda.shtml

Berger, Theodore. Quoted in "Restoring Memory, Repairing Damaged Brains: USC Viterbi biomedical engineers analyze -- and duplicate -- the neural mechanism of learning in rats." News Release. 17 June 2011 Los Angeles: USC Viterbi School of Engineering, 2011. Available from http://viterbi.usc.edu/news/news/2011/restoring-memory-repairing.htm

Bering, Jesse. "The End?: Why So Many of Us Think Our Minds Continue On After We Die." *Scientific American Mind,* October/November 2008, p. 37.

Bicentennial Man. Directed by Chris Columbus. 2 hr. 12 min. Radiant Films, Columbia Pictures, Touchstone Pictures, 1492 Pictures, 1999. DVD Video.

Bierbower, Austin. "American Wastefulness." *Industrial America: The Environment and Social Problems, 1865-1920*. Compiled and edited by H. Wayne Morgan. Chicago: Rand McNally College Pub. Co., 1974.

Blakeslee, Sandra. "Monkey's Thoughts Propel Robot, A Step That May Help Humans." *New York Times* [online], 15 January 2008, available at http://www.nytimes.com/2008/01/15/science/15robo.html?_r=1&pagewanted=all

Blakeslee, Sandra. "Monkey's Thoughts Propel Robot, A Step That May Help Humans." *New York Times* [online], 15 January 2008, available at http://www.nytimes.com/2008/01/15/science/15robo.html?_r=1&pagewanted=all

Boia, Lucian. *Forever Young: A Cultural History of Longevity*. Translated by Trista Selous. London: Reaktion Books Ltd, 2004.

Borkat, Roberta Sarfatt. "Pride, Progress, and Swift's Struldbruggs." *Durham University Journal*, vol. 68, June 1976, pp. 126-134.

Bradstock, Ross A., Jann E. Williams and Malcolm A. Gill, ed. *Flammable Australia: The Fire Regimes and Biodiversity of a Continent*. Cambridge: Cambridge University Press, 2002.

Breaker Morant. Directed by Bruce Beresford. 1 hr. 47 min. Wellspring, 1979. DVD Video.

Brinkley, Alan. *American History: A Survey,* 11th Edition. New York: McGraw-Hill, 2003.

Bronowski, J. *The Ascent of Man.* Boston; Toronto: Little, Brown and Company, 1973.

Brooks, Rodney. "I, Rodney Brooks, Am a Robot," *IEEE Spectrum*, vol. 45, no. 6, June 2008, p. 69.

Brosnan, Sarah F., and Frans B. M. de Waal. "Monkeys Reject Unequal Pay." *Nature*, Vol. 425, pp. 297-299; September 18, 2003.

Brown, Kathryn. "A Radical Proposal." *Scientific American Presents*, vol. 11, no. 2, Summer 2000, p. 39.

Bryson, Bill. *A Short History of Nearly Everything.* New York: Broadway Books, 2003.

Bryson, Bill. *The Life and Times of the Thunderbolt Kid.* New York: Broadway Books, 2006.

Bryson, Bill. *The Lost Continent: Travels in Small Town America.* New York: Harper Perennial, 2001.

Burke, James. *The Day the Universe Changed.* Boston and Toronto: Little, Brown and Company, 1986.

Burke, James. *The Day the Universe Changed.* Directed by Richard Reisz. 9 hr. 10 min. BBC Production, 1985. Distributed by Ambrose Video Publishing, Inc. DVD Video.

Calvert, Clay."Freedom of Speech Extended to Corporations." T*he Encyclopedia of American Civil Liberties*, Volume 1, ed. Paul Finkelman. New York: Routledge.

Carroll, Rory. "Pope Says Sorry for Sins of Church." *The Guardian*, 13 March 2000, available at http://www.guardian.co.uk/world/2000/mar/13/catholicism.religion

Carroll, Sean B., Benjamin Prud'homme and Nicolas Gompel. "Regulating Evolution." *Scientific American*, vol.298, pp. 60-67; May 2008.

Carwardine, Mark, Erich Hoyt, R. Ewan Fordyce, and Peter Gill. *Whales, Dolphins & Porpoises.* Alexandria, VA: Nature Company: Time-Life Books, 1998.

Casciato, Tom. Producer. "Life on the Edge." *Now*, broadcast on KQED, March 29, 2002.

Cave, Stephen. *Immortality: The Quest to Live Forever and How it Drives Civilization.* New York: Crown Publishers, 2012.

CDC fact sheet on diphtheria, available from CDC website at http://www.cdc.gov/vaccines/pubs/pinkbook/downloads/dip.pdf

Chauvet, Jean-Marie, Éliette Brunel Deschamps, and Christian Hillaire. *Dawn of Art, The Chauvet Cave: The Oldest Known Paintings in the World.* New York: Harry N. Abrams, Inc., 1996.

Cocoon. Directed by Ron Howard. 1 hr. 57 min. 20th Century Fox, 1985, DVD Video.

Constitution of Ireland. Dublin: Government Publications Office, 1980.

Contact. Directed by Robert Zemeckis. 2 hr. 33 min. Warner Brothers, 1997, DVD Video.
Conwell, Russell Herman. *Acres of Diamonds*. New York and London: Harper & Brothers Publishers, 1915.
Crick, Francis. *The Astonishing Hypothesis: The Scientific Search for the Soul*. New York, London, Toronto, Sydney, Tokyo, Singapore: Simon & Shuster, 1994.
Dahm, Ralf. "Dying to See." *Scientific American*, vol.291, no. 4, October 2004, p. 82-89.
Dark Star. Directed by John Carpenter. 1 hr. 23 min. Jack H. Harris Enterprises; University of Southern California (USC), 1974, DVD Video.
Darwin, Charles. Letter to Hooker, J.D., 9 February 1865. Available at http://www.darwinproject.ac.uk/entry-4769
Darwin, Charles. *The Life and Letters of Charles Darwin*, vol. 1. Edited by Francis Darwin. New York: Basic Books, Inc., 1959.
Davies, Paul. "That Mysterious Flow." *Scientific American*, vol. 287, no. 3, September 2002, p. 47.
de Waal, Frans. "So Human, So Chimp." *The Human Spark*. Written and produced by Graham Chedd. Directed by Larry Engel.57 min. A co-production of Chedd-Angier-Lewis Productions and THIRTEEN in association with WNET.ORG, airdate: 13 January 2010.
Deacon, Terrence. *The Symbolic Species: The Co-evolution of Language and the Brain*. New York, London: W. W. Norton & Company, 1997.
Deretic, Vojo and Daniel J. Klionsky, "How Cells Clean House," *Scientific American*, vol. 298, no. 5, May 2008, pp. 74-81.
d'Errico, Francesco, Christopher Henshilwood, Marian Vanhaeren and Karen van Niekerk. "Nassarius Kraussianus Shell Beads from Blombos Cave: Evidence for Symbolic Behavior in the Middle Stone Age," *Journal of Human Evolution*, vol. 48, no. 1 (2005): 3-24.
Dezzani, Mark. "New Earthquake in Northern Italy Kills 16." *BBC News* [online], 29 May 2012. Available at http://www.bbc.co.uk/news/world-europe-18247659
Doidge, Normal. *The Brain That Changes Itself: Stories of Personal Triumph from the Frontiers of Brain Science*. New York: Viking, 2007.
Doyle, Rodger. "The Rich and Other Americans." *Scientific American*, vol.284, no. 2, February 2001, p. 26.
Doyle, Roger. "Religion in America", *Scientific American*, Vol. 288, #2, p. 26; February, 2003.
Dulbecco, Renato. *The Design of Life*. New Haven and London: Yale University Press, 1987.
Edelman, Gerry. "Changing Your Mind." *Scientific American Frontiers*. Written, produced and directed by Graham Chedd. 55 min. The Chedd-Angier Production Company, Inc., airdate: 21 November 2000.
Edles, Laura Desfor and Scott Appelrouth. *Sociological Theory in the Classical Era*. Thousand Oaks, California: Pine Forge Press, 2005.

Ehrlich, Paul R. and Anne H. Ehrlich. *The Population Explosion.* New York: Simon and Schuster, 1990.
Ehrlich, Paul R. *The Population Bomb.* New York: Ballantine Books Inc., 1971.
Eliade, Mircea. *From Primitives to Zen: A Thematic Sourcebook of the History of Religions.* New York and Evanston: Harper & Row, Publishers, 1967.
Eliot, T.S. "Little Gidding." *Collected Poems: 1909-1962.* San Diego; New York; London: Harcourt Brace Jovanovich, 1963.
Emerson, Ralph Waldo. *The American Scholar; Self-Reliance; Compensation.* New York, Cincinnati and Chicago: American Book Company, 1893.
Epicurus. *Epicurus, The Extant Remains / With Short Critical Apparatus, Translation and Notes by Cyril Bailey*, M.A. Translated by Cyril Bailey and Hermann Usener. Oxford: Clarendon Press, 1926.
Epicurus. *The Philosophy of Epicurus; Letters, doctrines, and parallel passages from Lucretius*; translated with commentary and an introductory essay on ancient materialism by George K. Strodach. Evanston, Ill: Northwestern University Press, 1963.
Ewald, Paul W. "The Evolution of Virulence." *Scientific American*, vol. 268, no. 4, April 1993, pp. 86-93.
Fern, Yvonne. *Gene Roddenberry: The Last Conversation.* Berkeley: University of California Press, 1994.
Frey, Jennifer. "Terri Schiavo's Unstudied Life – The Woman Who Is Now a Symbol And a Cause Hated the Spotlight." *Washington Post*, March 25, 2005, Sec. Style, p. C1.
George, Andrew, translator. *The Epic of Gilgamesh.* London: Penguin Books, 2003.
Ghent, William J. *Our Benevolent Feudalism.* New York: The Macmillan Company, 1902.
Gibbs, W. Wayt. "Synthetic Life." *Scientific American*, vol. 290, no. 5, May 2004, pp. 74-81.
Gill, Warren M. "Signal Considerations for Chronically Implanted Electrodes for Brain Interfacing." *Indwelling Neural Implants: Strategies for Contending with the In Vivo Environment.* Edited by Reichert WM. Boca Raton (FL): CRC Press, 2008.
Goodenough, Judith, Betty McGuire and Robert A. Wallace. *Perspectives on Animal Behavior.* New York: John Wiley, 2001.
Gorer, Geoffrey. "The Pornography of Death." This essay was first published in *Encounter* in October 1955, and was reprinted in Geoffrey Gorer, *Death, Grief, and Mourning.* New York: Doubleday, 1965.
Gorer, Geoffrey. *Death, Grief, and Mourning.* New York: Doubleday, 1965.
Gorst, Martin. *Measuring Eternity: The Search for the Beginning of Time.* New York: Broadway Books, 2001.
Grayson, Donald K. *The Establishment of Human Antiquity.* New York: Academic Press, 1983.
Greenemeier, Larry. "Monkey Think, Robot Do." *Scientific American* [online], 15 January 2008, available at http://www.scientificamerican.com/article.cfm?id=monkey-think-robot-do

Greenough, William T., John R. Larson and Ginger S. Withers. "Effects of Unilateral and Bilateral Training in a Reaching Task on Dendritic Branching of Neurons in the Rat Motor-Sensory Forelimb Cortex." *Behavioral and Neural Biology*, vol. 44, no. 2, September 1985, pp. 301-314.

Gruman, Gerald J. *A History of Ideas about the Prolongation of Life.* New York: Springer Publishing Company, 2003.

Hadingham, Evan. *Secrets of the Ice Age: The World of the Cave Artists.* New York: Walker, 1979.

Hale, Wayne, et al., (ed). *Wings in Orbit: Scientific and Engineering Legacies of the Space Shuttle.* Washington, D.C.: National Aeronautics and Space Administration, 2010.

Hassler, Susan. "Un-assuming The Singularity." *IEEE Spectrum*, vol. 45, no. 6, June 2008, p. 9.

Hawking, Stephen. "Did God Create the Universe?" *Curiosity*. Produced by Alan Eyres et al. 45 min. A production of Darlow Smithson Productions for Discovery Channel, airdate: 7 August 2011.

Hazen, Robert M. *Genesis: The Scientific Quest for Life's Origin.* Washington DC: Joseph Henry Press, 2005.

Hertz, John, Anders Krogh and Richard G. Palmer. *Introduction to the Theory of Neural Computation.* Redwood City, California: Addison-Wesley Publishing Company, 1991.

Herz, Rachel S. and Jonathan W. Schooler. "A Naturalistic Study of Autobiographical Memories Evoked by Olfactory and Visual Cues: Testing the Proustian Hypothesis." *The American Journal of Psychology*, vol. 115, no. 1 (Spring 2002), pp. 21-32.

Hofstadter, Richard. *Social Darwinism in American Thought.* Boston: Beacon Press, 1992.

Holt, Sarah. "The Man Who Couldn't Remember." *Nova ScienceNOW* [PBS Online]. Edited Susan K. Lewis. Posted 1 June 2009. Available from http://www.pbs.org/wgbh/nova/body/corkin-hm-memory.html

Hopkin, Karen. "Making Methuselah," *Scientific American Present: Your Bionic Future*. vol.10, no. 3, 1999, pp. 32-37.

Ingersoll, Robert. Quoted by Michael Shermer. *Why People Believe Weird Things: Pseudoscience, Superstition, and Other Confusions of Our Time.* New York: W.H. Freeman and Company, 1997.

Jackson, Danny P. *The Epic of Gilgamesh.* Introduction by Robert D. Biggs; an appreciation by James G. Keenan; illustrated by Thom Kapheim. Wauconda, Illinois: Bolchazy-Carducci Publishers, 2002.

James, P. D. Interview by Steve Paulson. "Elementary Holmes," *To the Best of Our Knowledge*. Wisconsin Public Radio, distributed the week of September 25, 2005, hour 1.

Johnson, Catherine. Promised Land or Purgatory. *Scientific American presents: The Quest to Beat Aging*, vol. 11, no. 2, Summer 2000, p. 96.

Johnson, Linda A. "KFC, McDonald's Fries and Chicken are Fattier in the U.S. than in Some Other Countries." *The Associated Press*, 12 April 2006.

Johnson, Paul. *The Civilization of Ancient Egypt.* New York: Atheneum, 1978.

Johnston, Alan. "Italy Quake Homeless in Emergency Shelters." *BBC News* [online], 21 May 2012. Available at http://www.bbc.co.uk/news/world-europe-18140543

Jones, Richard A.L. "Rupturing the Nanotech Rapture." *IEEE Spectrum*, vol. 45, no. 6, June 2008, pp. 64-67.

Juncosa, Barbara. "Is 100 the New 80?: Centenarians Studied to Find the Secret of Longevity." *Scientific American* [online], 28 October 2008, available at http://www.scientificamerican.com/article.cfm?id=centarians-studied-to-find-the-secret-of-longevity

Kelly, Florence. "Child Labor in Illinois," *Ninth Annual Convention of the International Association of Factory Inspectors of North America held at Providence, R. I., September 3-5, 1895.* Cleveland, Ohio: Forest City Printing House, 1895.

Kennedy, David M., Lizabeth Cohen, and Thomas A. Bailey. *The American Pageant: A History of the Republic.* New York and Boston: Houghton Mifflin Company, 2002.

Kennedy, John F. "Rice University, 12 September 1962." Digital identifier: USG-15-r29, [database online] Boston: John F. Kennedy Presidential Library and Museum. Available from http://www.jfklibrary.org/Asset-Viewer/MkATdOcdU06X5uNHbmqm1Q.aspx

Kennedy, John F. "Special Message to Congress on Urgent National Needs, 25 May 1961." *Papers of John F. Kennedy. Presidential Papers. President's Office Files.* [database online] Boston: John F. Kennedy Presidential Library and Museum. Available from http://www.jfklibrary.org/Asset-Viewer/Archives/JFKPOF-034-030.aspx

Kennedy, John F. "Toward a Strategy of Peace." *The Department of State Bulletin*, vol. 49, no. 1253, July 1, 1963, pp. 2-6.

Kennedy, Paul. *The Rise and Fall of the Great Powers: Economic Change and Military Conflict from 1500 to 2000.* New York: Random House, 1987.

Kenny, Enda. Quoted in: "This is a Republic, not the Vatican." *The Irish Times*, 21 July 2011. Available at http://www.irishtimes.com/newspaper/opinion/2011/0721/1224301061733.html

Kenny, Enda. Quoted in: Bob Simon, "The Archbishop of Dublin Challenges the Church." *60 Minutes*. Produced by Tom Anderson. 4 March 2012. Available at http://www.cbsnews.com/8301-18560_162-57390125/?tag=currentVideoInfo;videoMetaInfo

Kerry, John and Teresa Heinz Kerry. *This Moment on Earth: Today's New Environmentalists and Their Vision for the Future.* New York: Perseus, 2007.

Kiehl, Kent A. and Joshua W. Buckholtz. "Inside the Mind of a Psychopath." *Scientific American Mind*, vol. 21, no. 4, September/October 2010, pp. 22-29.

Kiehl, Kent A. and Joshua W. Buckholtz. "Inside the Mind of a Psychopath." *Scientific American Mind*. vol. 21, no. 4, September/October 2010, pp. 22-29.

Kipling, Rudyard. "The Last Department." *Departmental Ditties and Ballads and Barrack Room Ballads.* New York: Doubleday, Page & Company, 1915.

Klein, Maury. *The Genesis of Industrial America, 1870-1920.* Cambridge; New York: Cambridge University Press, 2007.

Knight, Matthew. "Mapping Out a New Era in Brain Research." *CNN* [online], 9 April 2012. Available at http://www.cnn.com/2012/03/01/tech/innovation/brain-map-connectome/index.html

Knight, Robert. Quoted by Jason Palmer. "Science Decodes 'Internal Voices.'" *BBC News* [Online], 31 January 2012, available at http://www.bbc.co.uk/news/science-environment-16811042

Koch, Christof and Giulio Tononi. "A Test for Consciousness" *Scientific American*, vol. 304, no. 6, June 2011, pp. 44-47.

Koch, Christof and Giulio Tononi. "Can Machines be Conscious?" *IEEE Spectrum*, vol. 45, no. 6, June 2008, pp. 55-59.

Koch, Christof. "Being John Malkovich." Scientific American Mind, vol. 22, no. 1, March/April 2011, pp. 18-19.

Koch, Christof. "Consciousness." A lecture presentation delivered by Christof Koch during the Cognitive Computing conference, hosted by the IBM Almaden Institute, 11 May, 2006. Video download available at http://www.almaden.ibm.com/institute/2006/agenda.shtml

Koch, Christof. *The Quest for Consciousness: A Neurobiological Approach.* Englewood, Colorado: Roberts and Company Publishers, 2004.

Kovacs, Maureen Gallery. *The Epic of Gilgamesh.* Stanford, California: Stanford University Press, 1989.

Kramer, Kenneth. *The Sacred Art of Dying: How World Religions Understand Death.* New York/Mahwah: Paulist Press, 1988.

Krauss, Lawrence M. and Robert J. Scherrer, "The End of Cosmology?" *Scientific American*, vol. 298, no. 3, March 2008, pp.46-53.

Kübler-Ross, Elizabeth. *Death is of Vital Importance: On Life, Death, and Life after Death* (New York: Station Hill Press, 1995).

Kübler-Ross, Elizabeth. *On Death and Dying.* New York: Macmillan Publishing Company, 1970.

Kurzweil, Ray and Terry Grossman, *Transcend: Nine Steps to Living Well Forever.* New York: Rodale, 2009.

Kurzweil, Ray. *The Age of Spiritual Machines: When Computers Exceed Human Intelligence.* New York: Penguin Books, 1999.

Kurzweil, Ray. *The Singularity is Near: When Humans Transcend Biology.* New York: Viking, 2005.

Landfield, Philip W., and Sam A. Deadwyler, ed. *Long-Term Potentiation: From Biophysics to Behavior.* New York: Alan R. Liss, Inc., 1988.

Larson, Edward J. and Larry Witham. "Leading Scientists Still Reject God." *Nature*, Vol. 394, p. 313; July 23, 1998.
Laudine, Catherine. *Aboriginal Environmental Knowledge: Rational Reverence.* Farnham, England; Burlington, VT: Ashgate, 2009.
Leakey, Richard E., and Roger Lewin. *Origins*. London: Macdonald and Jane's, 1979.
Leigh, Jason. "Can We Live Forever?" *NOVA ScienceNOW*. Written and produced by Vincent Liota. 55 min. Produced for WGBH/Boston by NOVA, airdate: 26 January 2011.
Leuba, James H. "Religious Beliefs of American Scientists." *Harper's*, Vol. 169 (1934): pp. 291-300.
Lewis, Julie. "Six Billion and Counting." *Scientific American*, vol. 283, no. 4, October 2000, pp. 30-32.
Lindenmayer, David and Mark Burgman. *Practical Conservation Biology*. Collingwood, Vic.: CSIRO Publishing, 2005.
Lucretius. *On the Nature of the Universe.* Translated by Ronald Melville. Oxford: Clarendon Press, 1997.
Luther, Martin. Quoted by Stephen Cave. *Immortality: The Quest to Live Forever and How it Drives Civilization.* New York: Crown Publishers, 2012.
Margulis, Lynn. *Symbiosis in Cell Evolution.* San Francisco: W. H. Freeman and Company, 1981.
Markram, Henry. "The Emergence of Intelligence in the Neocortical Microcircuit." A lecture presentation delivered by Henry Markram during the Cognitive Computing conference, hosted by the IBM Almaden Institute, 10 May, 2006. Video download available at http://www.almaden.ibm.com/institute/2006/agenda.shtml
Markram, Henry. Quoted in "'Blue Brain Founder Responds to Critics, Clarifies His Goals." *Science*, vol. 334, 11 November 2011, pp. 748-749.
Marshack, Alexander. *The Roots of Civilization*, New York: Moyer Bell Limited, 1991.
Mayr, Ernst and Carl Sagan. "The Search for Extraterrestrial Intelligence: Scientific Quest or Hopeful Folly?" *The Planetary Report*, vol. 16, no. 3, May/June 1996, pp. 4-13.
Mayr, Ernst. *What Evolution Is.* New York: Basic Books, 2001.
McCracken, Samuel. "Democratic Capitalism and the Standard of Living 1800-1980." *Modern Capitalism Volume 1: Capitalism and Equality in America.* Edited by Peter L. Berger. Lanham, New York and London: Hamilton Press, 1987.
McGrath, Matt. "Monkeys' Brain Waves Offer Quadriplegics Hope." *BBC News* [online], 6 October 2011. Available at http://www.bbc.co.uk/news/science-environment-15197124
McKay, Christopher P. "The Case for Mars." *The Planetary Report*, vol. 5, no. 2, March/April 1985, pp. 16-18.
Melton, Lisa. "AGE Breakers," *Scientific American*, vol. 283, no. 1, July 2000, p. 16.

Melton, Lisa. "Count to 10." *Scientific American*, vol. 286, no. 2, February 2002, p. 22.

Mirsky, Steve. "And the Winner Really Is...," *Scientific American*. vol.265, no. 2, August 2001, p. 95.

Mitchell, Stephen. *Gilgamesh: A New English Version*. New York, London, Toronto, Sydney: Free Press, 2004.

Moore, Gordon. "Tech Luminaries Address Singularity." *IEEE Spectrum Special Report: The Singularity* [online], June 2008, available at http://spectrum.ieee.org/computing/hardware/tech-luminaries-address-singularity

Morgan, H. Wayne. *Industrial America: The Environment and Social Problems, 1865-1920*. Compiled and edited by H. Wayne Morgan. Chicago: Rand McNally College Pub. Co., 1974.

Moyer Michael and Carina Storrs, "How Much is Left?" *Scientific American*, vol. 303, no. 3, September 2010, pp. 74-81.

Mozaffarian, Dariush et al. "Trans Fatty Acids and Cardiovascular Disease." *The New England Journal of Medicine*, vol. 354, 13 April 2006, pp. 1601-1613.

Muir, John. *John Muir: His Life and Letters and Other Writings*. Edited by Terry Gifford. London; Baton Wicks; Seattle: Mountaineers, 1996.

Muir, John. *Nature Writings*. Edited by William Cronon. New York: The Library of America: distributed to the trade in the U.S. by Penguin Books, USA, 1997.

Muir, John. Quoted by Linnie Marsh Wolfe. *Son of the Wilderness: The Life of John Muir*. Madison, Wisconsin: University of Wisconsin Press, 2003.

Muneoka, Ken, Manjong Han, and David M. Gardiner, "Regrowing Human Limbs." *Scientific American*, vol. 298, no. 4, April 2008, pp. 56-63.

Napier, Kristine. "Unproven Medical Treatments Lure Elderly," *FDA Consumer*. March 1994, pp. 33-37.

Nicolelis, Miguel. Interview by Charlie Rose, 16 October 2003, available at http://www.charlierose.com/guest/view/1771

Nimoy, Leonard. Interview by William Shatner. *Mind Meld: Secrets Behind the Voyage of a Lifetime*. DHG Productions in association with Melis Productions, executive producers William Shatner, Scott Zakarin, Rich Tackenberg, Peter Jaysen, directed by Peter Jaysen. 75 minutes. Razor Digital Entertainment, 2004. DVD Video.

Nordmann, Alfred. "Singular Simplicity." *IEEE Spectrum*, vol. 45, no. 6, June 2008, p. 61.

Nunn, Sam. "The Nuclear Chessboard, 2012." A discussion hosted by The Commonwealth Club of California, with George Shultz, William Perry, Sam Nunn and Philip Taubman at the Mark Hopkins Hotel in San Francisco, 23 February 2012.

Olshansky, S. Jay and Bruce A. Carnes, *The Quest for Immortality: Science at the Frontiers of Aging*. New York; London: W.W. Norton & Company: 2001.

Olshansky, S. Jay, and Bruce A. Carnes. *The Quest for Immortality: Science at the Frontiers of Aging.* New York, London: W.W. Norton & Company, 2001.

Panno, Joseph. *The Cell: Evolution of the First Organism.* New York: Facts on File, 2005.

Parker, Alice. Quoted in "Functioning Synapse Created Using Carbon Nanotubes." New Release. 21 April 2011 Los Angeles: USC Viterbi School of Engineering, 2011. Available at http://www.eurekalert.org/pub_releases/2011-04/uosc-rcf042111.php

Pasley, Brian. Quoted by Katie Moisse. "UC Berkeley Scientists Eavesdrop Inside the Mind." *ABC News* [Online], 31 January 2012, available at http://abcnews.go.com/Health/MindMoodNews/scientists-eavesdrop-inside-mind/story?id=15478684

Penrose, Roger. *The Emperor's New Mind: Concerning Computers, Minds, and the Laws of Physics.* London: Vintage, 1989.

Perry, William. "The Nuclear Chessboard, 2012." A discussion hosted by The Commonwealth Club of California, with George Shultz, William Perry, Sam Nunn and Philip Taubman at the Mark Hopkins Hotel in San Francisco, 23 February 2012.

Philipkoski, Kristen. "Ray Kurzweil's Plan: Never Die." *Wired Magazine* [online], 18 November 2002, available at http://www.wired.com/culture/lifestyle/news/2002/11/56448

Pinker, Steven. "Tech Luminaries Address Singularity." *IEEE Spectrum Special Report: The Singularity* [online], June 2008, available at http://spectrum.ieee.org/computing/hardware/tech-luminaries-address-singularity

Porter, Jennifer E. and Darcee L. McLaren, ed. *Star Trek and Sacred Ground: Explorations of Star Trek, Religion, and American Culture.* Albany, NY: State University of New York Press, 1999.

Poundstone, William. *Carl Sagan: A Life in the Cosmos.* New York: Henry Holt and Company, 1999.

Preston, John and Mark Bishop (eds). *Views into the Chinese Room: New Essays on Searle and Artificial Intelligence.* Oxford; New York: Clarendon Press, 2002.

Rask, Jon, et al. *Space Faring: The Radiation Challenge.* Publication: EP-2008-08-120-MSFC. Washington, D.C.: National Aeronautics and Space Administration, 2008.

Rawson, Hugh and Margaret Miner. *The Oxford Dictionary of American Quotations.* 2nd Edition. Oxford; New York: Oxford University Press, 2006.

Ray Bradbury's Martian Chronicles. Directed by Michael Andersen. 5 hr. 14 min. Charles Fries Productions, 1980. DVD Video.

Red Dwarf, Series 1, Episode 2: "Future Echoes." Written by Rob Grant & Doug Naylor, 1987. Shown on BBC television in February, 1988.

Remondino, P.C. "Longevity and Climate," *Minutes of the Twentieth Annual Session of the Medical Society of the State of California*. San Francisco: W.A. Woodward & Co, 1890.
Renner, Rebecca. "Scotchgard Scotched." *Scientific American*, vol. 284, no. 3, March 2001, p. 18.
Ritter, Malcolm. "Brain Electrodes Help Man Speak Again." *The Washington Post*, 1 August 2007.
Roig-Franzia, Manuel. "Long Legal Battle Over as Schiavo Dies: Florida Case Expected To Factor Into Laws For End-of-Life Rights." *Washington Post*, April 1, 2005, Sec. A, p. A1.
Rorabaugh, William J., Donald T. Critchlow, and Paula Baker. *America's Promise: A Concise History of the United States, Volume .1*. Lanham, Boulder, NewYork, Toronto and Oxford: Rowman & Littlefield Publishers, Inc., 2004.
Rose, Steven. "Human Agency in the Neurocentric Age." *EMBO Reports*, vol. 6, no. 11 (2005): pp. 1001-1005.
Rose, Steven. *The Future of the Brain: The Promise and Perils of Tomorrow's Neuroscience*. Oxford; New York: Oxford University Press, 2005.
Rothbard, Murray N. *A History of Money and Banking in the United States: The Colonial Era to World War II*. Edited by Joseph T. Salerno. Auburn, Alabama: Ludwig von Mises Institute, 2002.
Rumsfeld, Donald. "DoD News Briefing – Secretary Rumsfeld and Gen. Myers," News Transcript, 12 Feb 2002. Washington, D.C.: Federal News Service, Inc., 2002. Available from http://www.defense.gov/transcripts/transcript.aspx?transcriptid=2636
Sacks, Oliver. *The Man Who Mistook His Wife for a Hat*. London: Picador, 1986.
Sagan, Carl. "Let's Go to Mars Together." *The Planetary Report*, vol. 6, no. 4, July/August 1986, pp. 8-9.
Sagan, Carl. "The Shores of the Cosmic Ocean." *Cosmos*. Directed by Adrian Malone. 1 hr. Cosmos Studios, 1980, DVD Video.
Sagan, Carl. *Billions and Billions: Thoughts on Life and Death at the Brink of the Millennium*. New York: Random House, 1997.
Sagan, Carl. *Conversations with Carl Sagan*. Edited by Tom Head. Jackson, MS: University Press of Mississippi, 2006.
Sagan, Carl. *Cosmos*. London: Macdonald Futura Publishers, 1981.
Sagan, Carl. *Pale Blue Dot: A Vision of the Human Future in Space*. London: Headline, 1995.
Sagan, Carl. *The Demon Haunted World: Science as a Candle in the Dark*. London: Headline, 1996.
Sandars, N. K. *The Epic of Gilgamesh*. London: Penguin Books, 1977.
Savage-Rumbaugh, Sue, and Roger Lewin. *Kanzi: The Ape at the Brink of the Human Mind*. New York: John Wiley, 1994.
Schacter, Daniel. *The Seven Sins of Memory: How the Mind Forgets and Remembers*. Boston and New York: Houghton Mifflin Company, 2001.

Searle, John R. *The Mystery of Consciousness*. New York: The New York Review of Books, 1997.
Searle, John. "'I Married a Computer': An Exchange." Searle replies to Kurzweil's response to a review of *The Age of Spiritual Machines: When Computers Exceed Human Intelligence*, by Ray Kurzweil, *The New York Review of Books*, 20 May 1999.
Searle, John. "Beyond Dualism." A lecture presentation delivered by John Searle during the Cognitive Computing conference, hosted by the IBM Almaden Institute, 11 May, 2006. Video download available at http://www.almaden.ibm.com/institute/2006/agenda.shtml
Searle, John. "I Married a Computer." Review of *The Age of Spiritual Machines: When Computers Exceed Human Intelligence*, by Ray Kurzweil. *The New York Review of Books*. 8 April 1999.
Searle, John. "What is the Nature of Personal Identity?" *Closer to Truth* website. Available at http://www.closertotruth.com/video-profile/What-is-the-Nature-of-Personal-Identity-John-Searle-/439
Sejnowski, Terrence J. and Patric K. Stanton. "Covariance Storage in the Hippocampus." *An Introduction to Neural and Electronic Networks*, Second Edition. Edited by Steven F. Zornetzer et al. San Diego: Academic Press, 1995.
Seung, Sebastian. *Connectome: How the Brain's Wiring Makes Us Who We Are*. Boston; New York: Houghton Mifflin Harcourt, 2012.
Shakespeare, William. *The Tragedy of Hamlet, Prince of Denmark*. ed. Sylvan Barnet. 2nd Revised Edition. New York: Penguin Group, 1998.
Shankar, Anoop, Jie Xiao and Alan Ducatman. "Perfluoroalkyl Chemicals and Chronic Kidney Disease in US Adults." *American Journal of Epidemiology*, vol. 174, no. 8, 26 August 2011, pp. 893-900.
Shermer, Michael. "The Shamans of Scientism." *Scientific American*, vol. 286, no. 6, June 2002, p. 35.
Shostak, Seth, and Alex Barnett. *Cosmic Company: The Search for Life in the Universe*. Cambridge: Cambridge University Press, 2003.
Shostak, Seth. *Sharing the Universe: Perspectives on Extraterrestrial Life*. Berkeley: Berkeley Hills Books, 1998.
Shreeve, James. "The Mind is what the Brain Does." *National Geographic*, vol. 207, no. 3, March 2005, pp. 2-31.
Smith, Crosbie and M. Norton Wise. *Energy and Empire: A Biographical Study of Lord Kelvin*. Cambridge; New York: Cambridge University Press, 1989.
Spencer, A.J. *Death in Ancient Egypt*. Harmondsworth, Middlesex, England; New York: Penguin Books, 1982.
Spencer, Herbert. *Social Statics; together with Man Versus the State*. New York and London: D. Appleton and Company, 1892.
Spiro, Melford E. "Religion and the Irrational." *Symposium on New Approaches to the Study of Religion*. Edited by June Helm. Seattle: University of Washington Press, 1964.

Srivastava, Pramod K. "New Jobs for Ancient Chaperones." *Scientific American*, vol. 299, no. 1, July 2008, pp. 50-55.

Stanner, W. E. H. "The Dreaming." *Traditional Aboriginal Society*, Second Edition. Edited by W.H. Edwards. South Yarra: Macmillan Education Australia Pty. Ltd., 1998.

Star Trek: Generations. Directed by David Carson. 1 hr. 58 min. Paramount, 1994. DVD Video.

Stark, Rodney and William Sims Bainbridge. *A Theory of Religion*. New Brunswick, New Jersey: Rutgers University Press, 1996.

Stark, Rodney. "Age and Faith: A Changing Outlook or an Old Process?" *Sociological Analysis*, vol. 29 (1968), p. 1-10.

Stender, Steen, Jorn Dyerberg and Arne Astrup. "High Levels of Industrially Produced Trans Fat in Popular Fast Foods." *The New England Journal of Medicine*, vol. 354, 13 April 2006, pp. 1650-1652.

Sumner, William Graham. *Social Darwinism: Selected Essays of William Graham Sumner*. Englewood Cliffs, N. J.: Prentice-Hall, Inc., 1963.

Swift, Jonathan. *Gulliver's Travels*. New York: Random House, 1996.

Tattersall, Ian. "How We Came to be Human." *Scientific American*, vol. 285, no. 6, December 2001, pp. 56-63.

Tattersall, Ian. *Becoming Human: Evolution and Human Uniqueness*. New York, San Diego, London: Harcourt Brace & Company, 1998.

Taylor, John H. *Death and the Afterlife in Ancient Egypt*. Chicago: The University of Chicago Press, 2001.

Taylor, Thomas N., Edith L. Taylor, and Michael Krings. *Paleobotany: The Biology and Evolution of Fossil Plants*. Amsterdam; Boston: Academic Press, 2009.

The Six Million Dollar Man: The Complete First Season. Produced by Harve Bennett. 14 hr. 42 min. Universal Studios, 1974, DVD Video.

Thimmesh, Catherine. *Team Moon: How 400,000 People Landed Apollo 11 on the Moon*. New York: Houghton Mifflin, 2006.

Tipler, Frank. "How Significant is an Expanding Universe?" *Closer to Truth* website. Available at http://www.closertotruth.com/video-profile/How-Significant-is-an-Expanding-Universe-Frank-Tipler-/1036

Tipler, Frank. *The Physics of Immortality: Modern Cosmology, God and the Resurrection of the Dead*. London: MacMillan, 1994.

Toynbee, Arnold Joseph. *Life after Death*. New York, St. Louis, and San Francisco: McGraw-Hill Book Company, 1976.

Tyson, Neil deGrasse. "Can We Live Forever?" *NOVA ScienceNOW*. Written and produced by Vincent Liota. 55 min. Produced for WGBH/Boston by NOVA, airdate: 26 January 2011.

Tyson, Neil deGrasse. "Holy Wars: An Astrophysicist Ponders the God Question," *Science and Religion: Are They Compatible?* Edited by Paul Kurtz, Barry Karr and Ranjit Sandhu. New York: Prometheus Books, 2003.

Van Buren, Martin. Quoted by Howard Zinn, *A People's History of the United States*. New York: HarperPerennial, 1980.

Wachhorst, Wyn. "Carl Sagan, Visionary." *The Planetary Report*. 1997, no. 3, p. 22.

Walford, Roy. "Fat and Happy." *Scientific American Frontiers*. Written, produced and directed by John Angier, David Huntley and Andy Liebman. 55 min. The Chedd-Angier Production Company, Inc., airdate: 1 May 2001.

Waters, David. "Sorry Charlie: Church Apologizes to Darwin." *The Washington Post*, 17 September 2008, available at http://onfaith.washingtonpost.com/onfaith/undergod/2008/09/church_of_england_apologizes_t.html

Weinberg, Steven. "A Designer Universe." *Science and Religion: Are They Compatible?* Edited by Paul Kurtz, Barry Karr and Ranjit Sandhu. New York: Prometheus Books, 2003.

Weiten, Wayne. *Psychology: Themes & Variations*. Briefer Version. Fourth Edition. Pacific Grove, California: Wadsworth/Brooks/Cole, 2000).

Wenner, Melinda. "Regaining Lost Luster." *Scientific American*, vol. 298, no. 1, January 2008, pp. 18-20.

Whiten, Andrew, and Christophe Boesch. "The Culture of Chimpanzees." *Scientific American*, vol. 284, pp. 60-67; January 2001.

Willcox, Bradley et al. "FOXO3A Genotype is Strongly Associated with Human Longevity." *PNAS*, vol. 105, no. 37, 16 September 2008, pp. 13987-13992.

Willett, Walter C. and Meir J. Stampfer. "Rebuilding the Food Pyramid," *Scientific American*, vol. 288, no. 1, pp. 64-71.

Williams, Brian. *NBC Nightly News*. Airdate: 6 February 2012.

Wilson, Edward O. *The Future of Life*. New York: Alfred A. Knopf, 2002.

Witzel, L. "Behavior of the Dying Patient," *British Medical Journal*, 1975, 2, pp. 81-82.

Wolfram, Stephen. *A New Kind of Science*. Champaign, Illinois: Wolfram Media, 2002.

Wong, Kate. "The Morning of the Modern Mind." *Scientific American*, vol.292, pp. 86-95; June 2005.

Young, Liane et al. "Disruption of the right temporoparietal junction with transcranial magnetic stimulation reduces the role of beliefs in moral judgments." *PNAS*, vol. 107, no. 15, 13 April 2010, pp. 6753-6758.

Zak, Paul J. "The Neurobiology of Trust." *Scientific American*, vol. 298, no. 6, June 2008, pp. 88-95.

Zak, Paul. "The Neurobiology of Trust." *Scientific American*, vol. 298, no. 6, June 2008, p.95.

Zorpette, Glenn. "Waiting for the Rapture," *IEEE Spectrum*, vol. 45, no. 6, June 2008, pp. 34-35.

Zull, James E. *From Brain to Mind: Using Neuroscience to Guide Change in Education*. Sterling, Virginia: Stylus Publishing, LLC., 2011.

Zweibel, Ken, James Mason and Vasilis Fthenakis. "A Solar Grand Plan." *Scientific American*, vol. 298, no. 1, January 2008, pp. 64-73.

INDEX

Abri Blanchard, 14
Abydos, 32
Achilles, 40
Ackerman, Diane, 112, 135, 185
aging, 8, 49–50, 39, 41, 49–50, 202–8, 219
   age-related illnesses, 43
agriculture, 3–4, 160, 166
Akkadian, 5
Alcor, 149, 231, 233
Alda, Alan, 114–15, 206
Alexander the Great, 40
Alzheimer's disease, 37, 41, 145, 147
ancient Egypt, 10, 32–33, 38, 140
   mummies, 33
   shabtis, 33
androids, 87, 104, 106, 132
anesthesia, 8, 55
Aries, Philippe, 10, 20, 24–25
artificial brains, 89, 104, 112, 118, 120, 129, 132, 134–36, 140–42, 144, 148, 149, 200, 216, 219, 224–26
   ethics of, 144
artificial hearts, 129
Ashurbanipal, 4
asteroid impact, 71, 80, 159–61
astronomy, 49
   cosmology, 54, 156
   fate of the universe, 156
   the Big Bang, 54, 166
Attenborough, David, 15

aurignacian, 13–14
avatars, 209, 223
Bainbridge, William Sims, 15, 17–20, 26
Becker, Ernest, 51
belief in an afterlife, 2, 23
   Egyptian belief, 2, 10, 31–35
   heaven, 23, 26, 28, 121, 137, 148, 183–92, 212, 219, 220, 224, 230
   scientists, 19
Berger, Theodore, 103–4, 129, 145, 149
Bering, Jesse, 48, 230
Bicentennial Man, 50
Book of the Dead, 33
Boyne valley, 1, 3–4
Bracken, Jim, 63–65, 75, 84, 193
Bronowski, Jacob, 178–79
Brooks, Rodney, 67, 121, 180
Bryson, Bill, 44, 61, 74
Burke, James, 48, 79
caloric restriction, 205
cave paintings, 13
Cave, Stephen, 226
cellular biology, 10, 89
   cell division, 42
   cell nucleus, 92, 123
   DNA, 42, 60–61, 75, 89, 125–26, 164
   genes, 43, 89, 123–25, 164
   proteins, 89, 92, 123–26, 164, 203, 206
   RNA, 60, 125, 164
   telomeres, 42
Chaco Canyon, 2
Cherokees, 167
chromosomes, 42
civilization, 3–5, 7, 30–35, 56, 62, 69, 72, 75, 79, 108, 140, 153, 155, 158, 160–68, 173, 221–23
   and alien invasions, 161
   and climate change, 71, 78, 160, 169
   and consumerism, 63, 74, 76, 153
   and engineered viruses, 164
   and natural disasters, 162
   and nuclear weapons, 163

and population, 4, 19, 20, 31, 42, 50, 77–80, 138, 160, 168–69, 173, 179, 219
and progress, 138, 150
and the Holocaust, 167
and the Progressive Era, 74, 173
capitalism and, 64–69
corporations, 66, 78, 211
Industrial Revolution and, 67
industry, 66
longevity of, 79
progress, 62, 74
survival of, 71, 155, 170
survival of the fittest, 69
computers, 43, 76, 87–96, 100–127, 133–34, 142–57, 164, 209–18, 223, 232
and brains, 109, 112, 121
and Moore's Law, 149–50
and sentience, 109
digital information, 108, 210
HAL 9000, 105
processor architecture, 106
programming languages, 107
software, 43, 76, 88, 91, 106–13, 121, 150, 164, 185, 217
Watson, 127
consciousness, 9, 27, 35, 99–113, 118–27, 133–34, 144, 189, 210–16, 226, *See also* sentience
and complexity, 144
and intelligence, 134
and machines, 121, 147
and quantum mechanics, 122
Conwell, Russell, 69
Cosmos, 194
creationism, 189
Crick, Francis, 123, 125–26
Cro-Magnon, 12–14, 16, 24
Dark Star, 195
Darwin, Charles, 69, 155, 178
Deacon, Terrence, 12, 16–17, 48
death, 2–10, 16–39, 46–51, 49, 87, 126, 148, 153, 158, 165, 183, 190–91, 202–4, 207, 215, 219, 224, 234
apologism, 46–47, 49–50
attempts to cheat death, 39, *See also* quest for immortality

attitudes toward death, 20, 29, 30, 48, 51, 190, 230
childhood mortality, 44
death as a disease, 29
denial of death, 22, 148
digital information, 108–9, 214, 223–24
dinosaurs, 62, 71, 80–81, 91, 159, 160, 197
disease
    cancer, 38, 43, 126, 139, 151–52, 154, 199, 203–4, 226
    diphtheria, 153
    heart disease, 45, 171
    Lou Gehrig's disease, 206
    Parkinson's, 41, 146, 147
    polio, 44
    tuberculosis, 44
Easter Island, 55
Edison, Thomas, 152, 232
Enkidu, 5, 191
environmentalism, 76, 174, 184
Epicurus, 35, 38, 46
eternal life, 6, 36, 46, 47, *See also* quest for immortality
Ewald, Paul, 165
extraterrestrial intelligence, 29–30, 55–57, 80–81, 161–62, 193–96
Father Ted, 183
fMRI, 113–14, 119, 211, 213
Font de Gaume, 13
FOXO3A, 206
fractals, 90, 91
gene therapy, 126, 204
genetic code, 91
genetics, 123
Gilgamesh, 5–7, 10, 28, 37, 39, 46, 191, 227–28, 230–31, 233–34
Gorer, Geoffrey, 20–22, 20, 21, 22, 25
Granddad Waine, 8, 26–27, 65, 137, 157, 167, 187, 214, 221, 227
Gruman, Gerald, 45–46, 208, 241
Hamlet, 11
Hawking, Stephen, 120, 189
HDL, 171, 207
heat shock proteins, 206
Hiroshima, 162
human ancestry, 11–18, 24, 44, 61, 81

human brain, 7, 27, 37, 88–121, 127–34, 141–50, 200, 202, 210–17, 225–33
  amygdala, 99
  and brain scanners, 130, 133, 147, 213, 231, *See also* connectome
  and brain transplants, 130
  and moral judgement, 117
  and plasticity, 132
  and restoration of cognitive functions, 144
  architecture of, 97
  backup of, 211
  Cochlear implants, 145
  connectome, 128, 213–16, 225, 231, 233
  corpus callosum, 117
  DBS, 146
  development of, 89
  EEG, 114
  electrodes, 146
  evolution of, 135
  fusiform face area, 128
  hippocampus, 102–4, 129, 145, 149
  interfacing with, 115, 133, 146, 216
  lateral geniculate nucleus, 96
  left parahippocampal gyrus, 114
  middle temporal area, 97
  mind backups, 120
  plasticity of, 99
  psychopaths, 142
  right temporoparietal junction, 116–17
  thalamus, 95–96
  visual cortex, 96
  Wearing, Clive, 103
Human Connectome Project, 213
humanism, 55
humans
  and human achievement, 122
  and human atrocities, 167, 181
  and human identity, 103, 212, 226
  and self-awareness, 87, 91
  fabricating new body parts, 201
  health, 171
  mind reading, 119

nature of, 168
relationship with the natural world, 5, 11, 125, 169, 170–77, 205
spiritual existence, 15–16, 37, 137, 174, 176, 189, 192
symbolic thought, 13
the human mind, 37, *See also* consciousness
trust and trustworthiness, 142
what it means to be human, 11, 15, 168
ice age, 12–13
intelligence, 58, 61–62, 88, 105, 108, 112, 120, 126–35, 140–42, 195
   artificial, 126
   machine intelligence, 127
Kennedy, President John F, 71
Koch, Christof, 93–94, 96, 99–118, 121–22, 134, 144, 210, 230
Kübler-Ross, Elizabeth, 22–26, 208, 233
Kurzweil, Ray, 110, 121–28, 134–37, 147–49, 156–57, 220
Large Hadron Collider, 166
Lascaux, 14
Layard, Austen Henry, 4
life
   biochemistry, 10, 59–61
   enzymes, 59–60
   extraterrestrial life, 61, 195
   life expectancy, 44–45, 75, 205
   life on Earth, 58, 62
   life prolongation, 23, 71, 80, 199, 202, *See also* attempts to cheat death
   lifespan, 30, 41, 44–46, 80, 158, 202
   Miller-Urey Experiment, 60
   molecular biology, 58
Lord Kelvin, 156
Luther, Martin, 187
machines
   integration with, 135
Mars, 46, 151, 196–201
Martian Chronicles, 46, 197
medical science, 42, 44–45, 49, 98, 126, 146, 149, 151
Moon landing, 122, 139
Moore, Gordon, 149, 150
mortality, 3, 5, 9–10, 24–25, 45, 47, 48, 110, 155, 226, *See also* death
   Struldbruggs, 207–8
Muir, John, 175–77, 183, 189, 192, 224

mythology, 40
nanotechnology, 147–48
  nanobots, 147–48, 202, 212, 217
natural selection, 9, 11, 18, 43, 69, 71, 88, 189, 204
Neolithic, 1, 3
neurology, 37
neuroscience, 37, 89, 90–102, 113–17, 121, 129, 145–47, 211–12, 216
  cortical column, 113
  LTD, 102, 137, 141
  memory, 101
  neurotransmitters, 93, 94, 110, 121, 215
  oxytocin, 118, 142, 143
  synapses, 92–93, 102, 110, 212, 232–33
  synaptic patterns, 101, 103, 120, 131, 133, 212, 232
  TMS, 116–17
Newgrange, 1–4, 7, 19
Nicolelis, Miguel, 115, 216
Nimoy, Leonard, 7
Nineveh, 4
Nixon, President Richard, 126
nuclear weapons, 162–64
Osiris, 32
P.D. James, 30
Penrose, Roger, 121, 215
perfluoro-octanyl sulfonate, 172
Perry, William, 162
philosophy, 25, 35, 38, 46, 47, 64, 75, 104, 109, 128, 148, 151, 153, 220, 225, 226
  Epicurean, 35, 37–38
  Martian philosophy, 46
Plato, 38–39
Ponce de Leon, 39
prehistoric burial, 16
President Bush, 8
quest for immortality, 6, 27–28, 51, 104, 118, 123, 128, 141, 149, 227, 233
  cryonics, 231, 233
  fountain of youth, 39
  Guéniot, 41
  Herman Boerhaave, 40
  King David, 40

Voronoff, 41
religion, 10–27, 47–49, 177–90
   and Lent, 181–82
   and morality, 181–82, 186
   and the Catholic Church, 180, 185–86
   and the Pope, 178–79
   Calvinism, 174
   Christianity, 2, 20, 27, 32, 47, 53, 117, 178, 182, 187
   Christmas, 2, 142, 190
   faith, 23, 27, 53, 55, 61, 80, 149, 181, 183, 186, 191, 233
   Genesis, 54
   magic & superstition, 13, 15, 43, 59, 66, 149, 157
   mysticism, 24–25
   religious belief, 19
   science and, 19, 178, 185
   the human soul, 27–28, 33, 38, 47, 51
   theory of religion, 17
resurrection, 2, 188, 191, 232–33
robots, 50, 58, 83, 105, 110–16, 128
   Asimo, 105
   R2D2, 105
   Robonaut, 110
Rockefeller, John D, 68, 73
Sacks, Oliver, 37
Sagan, Carl, 49, 55–57, 80, 135, 166–70, 189–96, 201
San Bushmen, 14
science & technology, 166
science fiction, 47, 57, 83–88, 104–5, 118, 120, 128, 136, 143, 151, 152, 167, 194–95, 197, 199, 202, 214
Searle, John, 104, 109, 110, 112–13, 118, 126, 128, 134, 148, 226, 229, 233
sentience
   and machines, 87, 109, 136
SETI, 30, 56–58, 56, 62, *See also* extraterrestrial intelligence
Seung, Sebastian, 127–28, 127, 233
Shamhat, 5
Shostak, Seth, 29, 57
singularitarians, 148, 149, 217
Sophocles, 159
space exploration, 153, 196–201, 201, 219, 233
Spencer, Herbert, 68–69, 80

Spiro, Melford, 18
Star Trek, 9, 46, 57, 84–87, 91, 104, 131–32, 157, 193–95, 198, 215–16
    and transporter technology, 86
        Kirk, Captain James T, 193
        Lieutenant Commander Data, 104, 144
        Lieutenant Uhura, 131–32, 157, 198, 216
        McCoy, Dr. Leonard, 84, 131
        Picard, Captain Jean-Luc, 47, 87
        Scott, Montgomery, 86
Stark, Rodney, 15, 17–20, 26
stone age, 3–4
Sungir, Russia, 16
Superman, 209
supernatural, 10, 17–18, 26, 184, 185
Swift, Jonathan, 208
Tattersall, Ian, 12
television, 57, 75, 84–85, 108, 139, 140, 188, 194
The Six Million Dollar Man, 83
Tír na nÓg, 40, See also mythology
Toynbee, Arnold, 35, 38
transcontinental railroad, 66, 138, 139, 152
Turing, Alan, 111
Uruk, 4, 228
Vanderbilt, William, 73
Vatican, 8, 33, 181
virtual reality, 91, 217, 218–23
Walford, Roy, 205–6
warp drive, 57, 132, 161, 195
Watson, James, 123

Notes

Chapter 1

[1] According to Richard Leakey, at around two thousand years ago, about half of the human population was supported by agriculture. By the eighteenth century, he estimates that as few as ten percent retained a hunter-gatherer lifestyle. "During the first eight thousand years of agriculture the number of people on the planet rocketed thirty times, from ten million to three hundred million." Richard E. Leakey, *Origins* (London: Macdonald and Jane's, 1979), p. 176.
[2] Maureen Gallery Kovacs, translator, *The Epic of Gilgamesh* (Stanford, California: Stanford University Press, 1989), p. 3.
[3] Translation by E. A. Speiser, in *Ancient Near Eastern Texts* (Princeton, 1950), pp. 72-99, reprinted in Isaac Mendelsohn (ed.), *Religions of the Ancient Near East*, Library of Religion paperback series (New York, 1955), pp. 47-115, reprinted in Mircea Eliade, *From Primitives to Zen: A Thematic Sourcebook of the History of Religions* (New York and Evanston: 1967), p. 328.
[4] N. K. Sandars, translator, *The Epic of Gilgamesh* (London: Penguin Books, 1977), pp. 103-104.
[5] Stephen Mitchell, translator, *Gilgamesh: A New English Version* (New York, London, Toronto, Sydney: FREE PRESS, 2004), pp. 168-169.
[6] Andrew George, translator, *The Epic of Gilgamesh*, (London: Penguin Books, 2003), p. 151.
[7] Leonard Nimoy, quote from a recorded conversation between Nimoy and his colleague, William Shatner, *Mind Meld: Secrets Behind the Voyage of a Lifetime*, DHG Productions in association with Melis Productions, executive producers William Shatner, Scott Zakarin, Rich Tackenberg, Peter Jaysen, directed by Peter Jaysen, 75 minutes, Razor Digital Entertainment, 2004, DVD Video.
[8] Manuel Roig-Franzia, "Long Legal Battle Over as Schiavo Dies: Florida Case Expected To Factor Into Laws For End-of-Life Rights," *Washington Post*, April 1, 2005, Sec. A, p. A1.
Jennifer Frey, "Terri Schiavo's Unstudied Life – The Woman Who Is Now a Symbol And a Cause Hated the Spotlight," *Washington Post*, March 25, 2005, Sec. Style, p. C1.
[9] Mitchell, p. 178.
[10] Philippe Aries, *The Hour of Our Death*, First Vintage Books edition, translated by Helen Weaver (New York: Random House, Inc., 1982).
[11] Genesis 1:26, King James version of the Bible.
[12] William Shakespeare, *The Tragedy of Hamlet, Prince of Denmark*, ed. Sylvan Barnet, 2nd Revised Edition (New York: Penguin Group, 1998), p. 50.
[13] Sue Savage-Rumbaugh, and Roger Lewin, *Kanzi: The Ape at the Brink of the*

Human Mind (New York: John Wiley, 1994).
[14] Sarah F. Brosnan and Frans B. M. de Waal, "Monkeys Reject Unequal Pay," *Nature*, Vol. 425, September 18, 2003, pp. 297-299.
[15] Andrew Whiten and Christophe Boesch, "The Culture of Chimpanzees", *Scientific American*, vol. 284, January 2001, pp. 60-67.
[16] Terrence Deacon, *The Symbolic Species: The Co-evolution of Language and the Brain* (New York, London: W. W. Norton & Company, 1997), p. 22.
[17] Kate Wong, "The Morning of the Modern Mind," *Scientific American*, vol.292, June 2005, pp. 86-95.
[18] Ian Tattersall, *Becoming Human: Evolution and Human Uniqueness* (New York, San Diego, London: Harcourt Brace & Company, 1998), p. 7.
[19] Tattersall, pp. 7-8.
[20] Evan Hadingham, *Secrets of the Ice Age: The World of the Cave Artists* (New York: Walker, 1979), p. 70.
[21] Hadingham, pp. 201-202.
[22] Tattersall, p. 23.
[23] Tattersall, p. 19.
[24] Hadingham, p. 11
[25] Hadingham, p. 70.
[26] Alexander Marshack, *The Roots of Civilization* (New York: Moyer Bell Limited, 1991), p. 44-49.
[27] Tattersall, p. 196.
[28] David Attenborough, "Social Climbers," *Life of Mammals*, season 1, episode 9, produced by Mike Salisbury and Huw Cordey, 50 min., BBC, 2003, DVD Video.
[29] Tattersall, p. 11.
[30] Terrence Deacon, quote from "Have Humans Stopped Evolving?" a public lecture presentation delivered by Deacon during the Wonderfest event at Pimental Hall, UC Berkeley, 7 November, 2004.
[31] Deacon, *Symbolic Species*, p. 430.
[32] Tattersall, p. 216.
[33] Tattersall, p. 201.
[34] Rodney Stark and William Sims Bainbridge, *A Theory of Religion* (New Brunswick, New Jersey: Rutgers University Press, 1996), p. 39.
[35] Rodney Stark and William Sims Bainbridge, *A Theory of Religion* (New Brunswick, New Jersey: Rutgers University Press, 1996), p. 39.
[36] Stark and Bainbridge, pp. 38-39.
[37] Stark and Bainbridge, p. 39.
[38] Melford E. Spiro, "Religion and the Irrational," *Symposium on New Approaches to the Study of Religion*, edited by June Helm, (Seattle: University of Washington Press, 1964), p. 105.
[39] Stark and Bainbridge, p. 33.
[40] Stark and Bainbridge, p. 85.
[41] James H. Leuba, "Religious Beliefs of American Scientists," *Harper's*, Vol. 169 (1934): pp. 291-300.

[42] Edward J. Larson and Larry Witham, "Leading Scientists Still Reject God", *Nature*, Vol. 394, July 23, 1998, p. 313.
[43] Stark and Bainbridge, p. 50.
[44] Aries, p. 560.
[45] Geoffrey Gorer, "The Pornography of Death." This essay was first published in *Encounter* in October 1955, and was reprinted in Geoffrey Gorer, *Death, Grief, and Mourning* (New York: Doubleday, 1965), p. 195.
[46] Geoffrey Gorer, "Pornography of Death," pp. 196-197.
[47] Geoffrey Gorer, *Death, Grief, and Mourning* (New York: Doubleday, 1965), p. 127-128. Although his research was conducted in Britain, Gorer noted that the same trend in rejection of mourning had occurred in the US, and he believed that his data would have relevance here too.
[48] Gorer, *Death, Grief, and Mourning*, p. xxxii.
[49] Gorer, *Death, Grief, and Mourning*, p. ix-x.
[50] Aries, p. 10.
[51] Aries, p. 587.
[52] Aries, pp. 613-614.
[53] Elizabeth Kübler-Ross, M.D., *On Death and Dying* (New York: Macmillan Publishing Company, 1970).
[54] Excerpt from an interview for "Quality of Life: At the End of Life", a *Choices and Challenges* forum held March 24, 1994 at the Center for Interdisciplinary Studies, Virginia Tech, Blacksburg, Virginia 24061.
[55] Kübler-Ross, pp. 15.
[56] Rodney Stark, "Age and Faith: A Changing Outlook or an Old Process?" *Sociological Analysis*, vol. 29 (1968), p. 1-10.
[57] L. Witzel, "Behavior of the Dying Patient, *British Medical Journal*, 1975, 2, pp. 81-82.
[58] Kübler-Ross, p. 18.
[59] Elizabeth Kübler-Ross, M.D., *Death is of Vital Importance: On Life, Death, and Life after Death* (New York: Station Hill Press, 1995).
[60] Aries, p. 11.
[61] Excerpt from the BBC television show, *Red Dwarf*, Series 1, Episode 2: "Future Echoes", written by Rob Grant & Doug Naylor, 1987.
[62] Kübler-Ross, *Death and Dying*, p. 266.
[63] Stark and Bainbridge, p. 298.
[64] Mitchell, p. 159.

Chapter 2

[1] Meredith Halks-Miller, "Is There a Fountain of Youth for the Brain," a public lecture presentation delivered by Halks-Miller and Dale Bredesen during the Wonderfest 2004 event, at Pimental Hall, UC Berkeley, 7 November, 2004.

² P.D. James, interview by Steve Paulson, "Elementary Holmes," *To the Best of Our Knowledge*, Wisconsin Public Radio, distributed the week of September 25, 2005, hour 1.
³ Paul Johnson, *The Civilization of Ancient Egypt* (New York, Atheneum, 1978), p. 135.
⁴ Photo from Wikimedia Commons, at http://commons.wikimedia.org/wiki/File:Giseh_16.jpg.
⁵ Arnold Joseph Toynbee, *Life after Death* (New York, St. Louis, and San Francisco: McGraw-Hill Book Company, 1976), p. 6.
⁶Epicurus, *Epicurus, The Extant Remains / With Short Critical Apparatus, Translation and Notes by Cyril Bailey, M.A.*, trans. Cyril Bailey and Hermann Usener (Oxford: Clarendon Press, 1926), p. 85.
⁷ Lucretius, *On the Nature of the Universe*, trans. Ronald Melville (Oxford: Clarendon Press, 1997), p. 99.
⁸ Lucretius, p. 87.
⁹ Lucretius, pp. 96-97.
¹⁰ Translation by Jan Assmann, in *Death and Salvation in Ancient Egypt*, translated from the German by David Lorton (Cornell University Press, 2005), pp. 4-5.
¹¹ A popular saying based on a conflation of two biblical quotes from Ecclesiastes (8:15) and Isaiah (22:13).
¹² Oliver Sacks, *The Man Who Mistook His Wife for a Hat* (London: Picador, 1986), p. x.
¹³ *Breaker Morant*, directed by Bruce Beresford, 1 hr. 47 min., Wellspring, 1979, DVD Video.
¹⁴ Toynbee, p. 7.
¹⁵ Lucian Boia, *Forever Young: A Cultural History of Longevity*, translated by Trista Selous (London: Reaktion Books Ltd, 2004), pp. 18-19.
¹⁶ Gerald J. Gruman, *A History of Ideas about the Prolongation of Life* (New York: Springer Publishing Company, 2003), p. 39.
¹⁷ P.C. Remondino, "Longevity and Climate," *Minutes of the Twentieth Annual Session of the Medical Society of the State of California*, (San Francisco: W.A. Woodward & Co, 1890), p. 284.
¹⁸ Boia, p. 139.
¹⁹ Robin Marantz Henig, "Living Longer: What Really Works?" *Scientific American presents: The Quest to Beat Aging*, vol. 11, no. 2, Summer 2000, p. 36.
²⁰ Kristine Napier, "Unproven Medical Treatments Lure Elderly," *FDA Consumer*, March 1994, pp. 33-37.
²¹ Fountain of Youth, by Lucas Cranach the Elder. From the Wikimedia Commons, at http://en.wikipedia.org/wiki/File:Lucas_Cranach_d._%C3%84._007.jpg
²² Bill Bryson, *The Lost Continent: Travels in Small Town America* (New York: Harper Perennial, 2001), p. 164.
²³ William J. Rorabaugh, Donald T. Critchlow, and Paula Baker, *America's*

Promise: A Concise History of the United States, Volume 1 (Lanham, Boulder, New York, Toronto and Oxford: Rowman & Littlefield Publishers, Inc., 2004), p. 47.

[24] Karen Hopkin, "Making Methuselah," *Scientific American Presents: Your Bionic Future*, vol.10, no. 3, 1999, pp. 32-37.

[25] Quoted by Steve Mirsky, "And the Winner Really Is...," *Scientific American*, vol.265, no. 2, August 2001, p. 95.

[26] Quoted by Hopkin, p. 34.

[27] Gruman, p. 4.

[28] *Ray Bradbury's Martian Chronicles*, directed by Michael Andersen, 5 hr. 14 min., Charles Fries Productions, 1980, DVD Video.

[29] *Star Trek: Generations*, directed by David Carson, 1 hr. 58 min., Paramount, 1994, DVD Video.

[30] John 3:16, King James Version of the Bible.

[31] Jesse Bering, "The End?: Why So Many of Us Think Our Minds Continue On After We Die," *Scientific American Mind*, October/November 2008, p. 36.

[32] Quoted in Bering, p. 37.

[33] Terrence Deacon, *The Symbolic Species: The Co-evolution of Language and the Brain* (New York, London: W. W. Norton & Company, 1997), p. 437.

[34] Deacon, p. 437.

[35] James Burke, *The Day the Universe Changed* (Boston and Toronto: Little, Brown and Company, 1986), p. 28.

[36] Wyn Wachhorst, "Carl Sagan, Visionary," *The Planetary Report*, 1997, no. 3, p. 22.

[37] Catherine Johnson, Promised Land or Purgatory, *Scientific American presents: The Quest to Beat Aging*, vol. 11, no. 2, Summer 2000, p. 96.

[38] *Bicentennial Man*, directed by Chris Columbus, 2 hr. 12 min., Radiant Films, Columbia Pictures, Touchstone Pictures, 1492 Pictures, 1999, DVD Video.

[39] Ernest Becker, *The Denial of Death* (New York: Simon & Schuster, 1997), pp. 11-12.

[40] *Babe*, directed by Chris Noonan, 1 hr. 34 min., Kennedy Miller Films, 1995, DVD Video.

Chapter 3

[1] *Contact*, directed by Robert Zemeckis, 2 hr. 33 min., Warner Brothers, 1997, DVD Video.

[2] Lisa Melton, "Count to 10," *Scientific American*, vol. 286, no. 2, February 2002, p. 22.

[3] William Poundstone, *Carl Sagan: A Life in the Cosmos* (New York: Henry Holt and Company, 1999), p. 190.

[4] Seth Shostak, *Sharing the Universe: Perspectives on Extraterrestrial Life* (Berkeley: Berkeley Hills Books, 1998), p. 146.

[5] *Cocoon*, directed by Ron Howard, 1 hr. 57 min., 20th Century Fox, 1985, DVD Video.
[6] Ernst Mayr and Carl Sagan, "The Search for Extraterrestrial Intelligence: Scientific Quest or Hopeful Folly?" *The Planetary Report*, vol. 16, no. 3, May/June 1996, pp. 4-13.
[7] Thomas N. Taylor, Edith L. Taylor, and Michael Krings, *Paleobotany: The Biology and Evolution of Fossil Plants* (Amsterdam; Boston: Academic Press, 2009), p.47.
[8] Photo from the website of the National Astronomy and Ionosphere Center, Arecibo Observatory, Puerto Rico, a facility of the NSF operated by Cornell University, at http://www.naic.edu
[9] Renato Dulbecco, *The Design of Life* (New Haven and London: Yale University Press, 1987), p.30.
[10] Bill Bryson, *A Short History of Nearly Everything* (New York: Broadway Books, 2003), p. 288.
[11] Lynn Margulis, *Symbiosis in Cell Evolution* (San Francisco: W. H. Freeman and Company, 1981), p.80.
[12] Mayr and Sagan, p.4.
[13] "There's a Great Big Beautiful Tomorrow," Original theme song for Walt Disney's "Carousel of Progress," written by Robert B. Sherman and Richard M. Sherman, published in 1964.
[14] Clay Calvert, "Freedom of Speech Extended to Corporations," T*he Encyclopedia of American Civil Liberties*, Volume 1, ed. Paul Finkelman (New York: Routledge), p. 650.
[15] Quoted in Hugh Rawson and Margaret Miner, *The Oxford Dictionary of American Quotations,* 2nd Edition (Oxford; New York: Oxford University Press, 2006), p. 91.
[16] Alan Brinkley, *American History: A Survey,* 11th Edition (New York: McGraw-Hill, 2003), p. 490.
[17] Maury Klein, *The Genesis of Industrial America, 1870-1920* (Cambridge; New York: Cambridge University Press, 2007), p.140.
[18] "Unsatisfactory Conditions Found in Syracuse Plants," *The Post-Standard*, Syracuse, New York, November 30, 1911.
[19] Quoted in "Fourth Report of the New York State Factory Investigating Commission," *Documents of the Senate of the State of New York Legislature: One Hundred and Thirty-Eighth Session*, vol. 14, no. 43, part 5, Transmitted to the legislature, February 15, 1915 (Albany, New York: J. B. Lyon Company, Printers, 1915), p. 2810.
[20] Florence Kelly, "Child Labor in Illinois," *Ninth Annual Convention of the International Association of Factory Inspectors of North America held at Providence, R. I., September 3-5, 1895* (Cleveland, Ohio: Forest City Printing House, 1895), p. 94.
[21] Quoted in Brinkley, p. 484.
[22] Herbert Spencer, *Social Statics; together with Man Versus the State* (New York and London: D. Appleton and Company, 1892), p. 206.

[23] Russell Herman Conwell, *Acres of Diamonds* (New York and London: Harper & Brothers Publishers, 1915), p. 21.
[24] William Graham Sumner, *Social Darwinism: Selected Essays of William Graham Sumner* (Englewood Cliffs, N. J.: Prentice-Hall, Inc., 1963), p. 14.
[25] Sumner, p. 157.
[26] Photo from Library of Congress, at http://www.loc.gov/pictures/resource/nclc.01581/
[27] President Kennedy, "Toward a Strategy of Peace," *The Department of State Bulletin*, vol. 49, no. 1253, July 1, 1963, pp. 2-6.
[28] Tom Casciato, producer, "Life on the Edge," *Now*, broadcast on KQED, March 29, 2002.
[29] Sumner, p. 17.
[30] Laura Desfor Edles and Scott Appelrouth, *Sociological Theory in the Classical Era* (Thousand Oaks, California: Pine Forge Press, 2005), p.84.
[31] Ian Tattersall, "How We Came to be Human" *Scientific American*, vol. 285, no. 6, December 2001, pp. 56-63.
[32] Quoted in William J, Ghent, *Our Benevolent Feudalism* (New York: The Macmillan Company, 1902), p. 29.
[33] Klein, p.111.
[34] Quoted in Klein, p.63.
[35] Bill Bryson, *The Life and Times of the Thunderbolt Kid* (New York: Broadway Books, 2006), p. 6.
[36] Rodger Doyle, "The Rich and Other Americans" *Scientific American*, vol.284, no. 2, February 2001, p. 26.
[37] Austin Bierbower, "American Wastefulness," *Industrial America: The Environment and Social Problems, 1865-1920*, compiled and edited by H. Wayne Morgan (Chicago: Rand McNally College Pub. Co., 1974), p. 105.
[38] Edward O. Wilson, *The Future of Life* (New York: Alfred A. Knopf, 2002), p. 23.
[39] Ralph Waldo Emerson, *The American Scholar; Self-Reliance; Compensation* (New York, Cincinnati and Chicago: American Book Company, 1893), p. 45.
[40] Quoted in Rawson and Miner, p. 27.
[41] Paul R. Ehrlich, *The Population Bomb* (New York: Ballantine Books Inc., 1971), p. 7.
[42] Paul R. Ehrlich and Anne H. Ehrlich, *The Population Explosion* (New York: Simon and Schuster, 1990), p. 211.
[43] Albert A. Bartlett, "Forgotten Fundamentals of the Energy Crisis," *American Journal of Physics*, vol. 46, no. 9, September 1978, pp. 876-888.
[44] James Burke, *The Day the Universe Changed*, directed by Richard Reisz, 9 hr. 10 min., BBC Production, 1985, distributed by Ambrose Video Publishing, Inc., DVD Video.
[45] W. E. H. Stanner, "The Dreaming," *Traditional Aboriginal Society*, Second Edition, edited by W.H. Edwards (South Yarra: Macmillan Education Australia Pty. Ltd., 1998), p. 227.
[46] Catherine Laudine, *Aboriginal Environmental Knowledge: Rational*

Reverence (Farnham, England; Burlington, VT: Ashgate, 2009), p. 78.
[47] "Chemical Controls," *Scientific American*, vol. 302, no. 4, April, 2010, p. 30.
[48] Carl Sagan, *Cosmos* (London: Macdonald Futura Publishers, 1981), p. 284.
[49] Julie Lewis, "Six Billion and Counting" *Scientific American*, vol. 283, no. 4, October 2000, pp. 30-32.

Chapter 4

[1] *The Six Million Dollar Man: The Complete First Season*, produced by Harve Bennett, 14 hr. 42 min., Universal Studios, 1974, DVD Video.
[2] "Relics," *Star Trek: The Next Generation*, season 6, episode 4, directed by Alexander Singer, 46 min., Paramount Television, 1992, DVD Video.
[3] "Mudd's Women," *Star Trek: The Original Series*, season 1, episode 6, directed by Harvey Hart, 50 min., Desilu Productions, 1966, DVD Video.
[4] "Second Chances," *Star Trek: The Next Generation*, season 6, episode 24, directed by LeVar Burton, 46 min., Paramount Television, 1993, DVD Video.
[5] "The Neutral Zone," *Star Trek: The Next Generation*, season 1, episode 26, directed by James L. Conway, 46 min., Paramount Television, 1988, DVD Video.
[6] "The Measure of a Man," *Star Trek: The Next Generation*, season 2, episode 9, directed by Robert Scheerer, 46 min., Paramount Television, 1989, DVD Video.
[7] Stephen Wolfram, *A New Kind of Science* (Champaign, Illinois: Wolfram Media, 2002).
[8] Craig Reynolds's Boids demonstration can be found at http://www.red3d.com/cwr/boids/
[9] James Shreeve, "The Mind is what the Brain Does" *National Geographic*, vol. 207, no. 3, March 2005, pp. 2-31.
[10] Photo from Wikimedia Commons, at http://commons.wikimedia.org/wiki/File:Dendrite_(PSF).png
[11] Christof Koch, *The Quest for Consciousness: A Neurobiological Approach* (Englewood, Colorado: Roberts and Company Publishers, 2004), p. 74.
[12] Judith Goodenough, Betty McGuire and Robert A. Wallace, *Perspectives on Animal Behavior* (New York: John Wiley, 2001), p. 107.
[13] Koch, p. 38.
[14] Koch, p. 302.
[15] Koch, p. 59.
[16] "Changing Your Mind," *Scientific American Frontiers*, written, produced and directed by Graham Chedd, 55 min., The Chedd-Angier Production Company, Inc., airdate: 21 November 2000.
[17] Shreeve, p. 21.
[18] Koch, pp. 305-306.
[19] Christof Koch and Giulio Tononi, "Can Machines be Conscious?" *IEEE Spectrum*, vol. 45, no. 6, June 2008, pp. 55-59.

20 Christof Koch, "Being John Malkovich," Scientific American Mind, vol. 22, no. 1, March/April 2011, pp. 18-19.
21 Koch, *The Quest for Consciousness*, pp. 188-189.
22 Quoted in Normal Doidge, *The Brain That Changes Itself: Stories of Personal Triumph from the Frontiers of Brain Science* (New York: Viking, 2007), p. 63.
23 Terrence J. Sejnowski and Patric K. Stanton, "Covariance Storage in the Hippocampus," *An Introduction to Neural and Electronic Networks*, Second Edition, ed. Steven F. Zornetzer et al. (San Diego: Academic Press, 1995), pp. 391-403.
24 William T. Greenough, John R. Larson and Ginger S. Withers, "Effects of Unilateral and Bilateral Training in a Reaching Task on Dendritic Branching of Neurons in the Rat Motor-Sensory Forelimb Cortex," *Behavioral and Neural Biology*, vol. 44, no. 2, September 1985, pp. 301-314.
25 Rachel S. Herz and Jonathan W. Schooler, "A Naturalistic Study of Autobiographical Memories Evoked by Olfactory and Visual Cues: Testing the Proustian Hypothesis," *The American Journal of Psychology*, vol. 115, no. 1 (Spring 2002), pp. 21-32.
26 John Searle, "Beyond Dualism," a lecture presentation delivered by John Searle during the Cognitive Computing conference, hosted by the IBM Almaden Institute, 11 May, 2006. Video download available at http://www.almaden.ibm.com/institute/2006/agenda.shtml

Chapter 5

1 Ray Kurzweil, *The Singularity is Near: When Humans Transcend Biology* (New York: Viking, 2005), p. 200.
2 Ray Kurzweil, *The Age of Spiritual Machines: When Computers Exceed Human Intelligence* (New York: Penguin Books, 1999), p. 129.
3 Christof Koch and Giulio Tononi, "A Test for Consciousness" *Scientific American*, vol. 304, no. 6, June 2011, pp. 44-47.
4 Diane Ackerman, *An Alchemy of the Mind: The Marvel and Mystery of the Brain* (New York: Scribner, 2004), p. 54.
5 "IBM Unveils Cognitive Computing Chips," News Release, 18 August 2011 (Armonk, New York: IBM Media Relations, 2011)
6 Henry Markram, "The Emergence of Intelligence in the Neocortical Microcircuit," a lecture presentation delivered by Henry Markram during the Cognitive Computing conference, hosted by the IBM Almaden Institute, 10 May, 2006. Video download available at http://www.almaden.ibm.com/institute/2006/agenda.shtml
7 Daniel Schacter, *The Seven Sins of Memory: How the Mind Forgets and Remembers* (Boston and New York: Houghton Mifflin Company, 2001), p. 26.

[8] Miguel Nicolelis, interview by Charlie Rose, 16 October 2003, available at http://www.charlierose.com/guest/view/1771

[9] Larry Greenemeier, "Monkey Think, Robot Do," *Scientific American* [online], 15 January 2008, available at http://www.scientificamerican.com/article.cfm?id=monkey-think-robot-do

[10] Sandra Blakeslee, "Monkey's Thoughts Propel Robot, A Step That May Help Humans," *New York Times* [online], 15 January 2008, available at http://www.nytimes.com/2008/01/15/science/15robo.html?_r=1&pagewanted=all

[11] Nicolelis (2003).

[12] Liane Young et al., "Disruption of the right temporoparietal junction with transcranial magnetic stimulation reduces the role of beliefs in moral judgments," *PNAS*, vol. 107, no. 15, 13 April 2010, pp. 6753-6758.

[13] Christof Koch, *The Quest for Consciousness: A Neurobiological Approach* (Englewood, Colorado: Roberts and Company Publishers, 2004), p. 291.

[14] Paul J. Zak, "The Neurobiology of Trust," *Scientific American*, vol. 298, no. 6, June 2008, pp. 88-95.

[15] Gerry Edelman, "Changing Your Mind," *Scientific American Frontiers*, written, produced and directed by Graham Chedd, 55 min., The Chedd-Angier Production Company, Inc., airdate: 21 November 2000.

[16] John Searle, "Beyond Dualism," a lecture presentation delivered by John Searle during the Cognitive Computing conference, hosted by the IBM Almaden Institute, 11 May, 2006. Video download available at http://www.almaden.ibm.com/institute/2006/agenda.shtml

[17] Photo by Shinji Nishimoto and Jack L. Gallant, UC Berkeley, 2011. Retrieved from Jack Gallant's website, at http://www.gallantlab.org/

[18] Jack L. Gallant et al., UC Berkeley, website can be found at http://www.gallantlab.org/

[19] Brian Pasley, Quoted by Katie Moisse, "UC Berkeley Scientists Eavesdrop Inside the Mind," *ABC News* [Online], 31 January 2012, available at http://abcnews.go.com/Health/MindMoodNews/scientists-eavesdrop-inside-mind/story?id=15478684

[20] Robert Knight, quoted by Jason Palmer, "Science Decodes 'Internal Voices'," *BBC News* [Online], 31 January 2012, available at http://www.bbc.co.uk/news/science-environment-16811042

[21] Steven Rose, "Human Agency in the Neurocentric Age," *EMBO Reports*, vol. 6, no. 11 (2005): pp. 1001-1005.

[22] Kurzweil, *The Singularity is Near*, p. 29.

[23] Rodney Brooks, "I, Rodney Brooks, Am a Robot," *IEEE Spectrum*, vol. 45, no. 6, June 2008, p. 69.

[24] Kurzweil, *The Singularity is Near*, p. 451.

[25] Christof Koch, "Consciousness," a lecture presentation delivered by Christof Koch during the Cognitive Computing conference, hosted by the IBM Almaden Institute, 11 May, 2006. Video download available at http://www.almaden.ibm.com/institute/2006/agenda.shtml

²⁶ Kurzweil, *The Age of Spiritual Machines*, p. 118.
²⁷ Christof Koch, quoted in Kurzweil, *The Singularity is Near*, p. 450.
²⁸ President Kennedy, "Special Message to Congress on Urgent National Needs, 25 May 1961," in *Papers of John F. Kennedy. Presidential Papers. President's Office Files*. [database online] (Boston: John F. Kennedy Presidential Library and Museum); available from http://www.jfklibrary.org/Asset-Viewer/Archives/JFKPOF-034-030.aspx
²⁹ President Kennedy, "Rice University, 12 September 1962," digital identifier: USG-15-r29, [database online] (Boston: John F. Kennedy Presidential Library and Museum); available from http://www.jfklibrary.org/Asset-Viewer/MkATdOcdU06X5uNHbmqm1Q.aspx
³⁰ Melinda Wenner, "Regaining Lost Luster," *Scientific American*, vol. 298, no. 1, January 2008, pp. 18-20.
³¹ Tim Beardsley, "Gene Therapy Setback," *Scientific American*, vol. 282, no. 2, February 2000, pp. 36-37.
³² Searle (2006).
³³ Susan Hassler, "Un-assuming The Singularity," *IEEE Spectrum*, vol. 45, no. 6, June 2008, p. 9.
³³ Kurzweil, *The Singularity is Near*, p. 29.
³⁴ Alice Parker, quoted in "Functioning Synapse Created Using Carbon Nanotubes," New Release, 21 April 2011 (Los Angeles: USC Viterbi School of Engineering, 2011), available at http://www.eurekalert.org/pub_releases/2011-04/uosc-rcf042111.php
³⁵ Steven Rose, *The Future of the Brain: The Promise and Perils of Tomorrow's Neuroscience*, (Oxford; New York: Oxford University Press, 2005), p. 217.
³⁶ "The Changeling," *Star Trek: The Original Series*, season 2, episode 3, directed by Marc Daniels, 50 min., Desilu Productions, 1967, DVD Video.
³⁷ "I, Mudd," *Star Trek: The Original Series*, season 2, episode 8, directed by Marc Daniels, 50 min., Desilu Productions, 1967, DVD Video.
³⁸ John Searle, "I Married a Computer," review of *The Age of Spiritual Machines: When Computers Exceed Human Intelligence*, by Ray Kurzweil, *The New York Review of Books*, 8 April 1999.
³⁹ Koch, "Consciousness" (2006).
⁴⁰ Ackerman, p. 93.
⁴¹ Carl Sagan, *Cosmos* (London: Macdonald Futura Publishers, 1981), p. 284.

Chapter 6

¹ Photo from Wikimedia Commons, at http://upload.wikimedia.org/wikipedia/commons/e/e4/1869-Golden_Spike.jpg
² Ray Kurzweil, *The Age of Spiritual Machines: When Computers Exceed Human Intelligence* (New York: Penguin Books, 1999), pp. 258-260.
³ Ray Kurzweil, *The Singularity is Near: When Humans Transcend Biology* (New York: Viking, 2005), p. 21.

[4] Kent A. Kiehl and Joshua W. Buckholtz, "Inside the Mind of a Psychopath," *Scientific American Mind*, vol. 21, no. 4, September/October 2010, pp. 22-29.
[5] Paul Zak, "The Neurobiology of Trust, *Scientific American*, vol. 298, no. 6, June 2008, p.95.
[6] Henry Markram, quoted in "'Blue Brain Founder Responds to Critics, Clarifies His Goals," *Science*, vol. 334, 11 November 2011, pp. 748-749.
[7] Christof Koch, "Consciousness," a lecture presentation delivered by Christof Koch during the Cognitive Computing conference, hosted by the IBM Almaden Institute, 11 May, 2006. Video download available at http://www.almaden.ibm.com/institute/2006/agenda.shtml
[8] Theodore Berger, quoted in "Restoring Memory, Repairing Damaged Brains: USC Viterbi biomedical engineers analyze -- and duplicate -- the neural mechanism of learning in rats," News Release, 17 June 2011 (Los Angeles: USC Viterbi School of Engineering, 2011), available from http://viterbi.usc.edu/news/news/2011/restoring-memory-repairing.htm
[9] Theodore Berger, from a panel discussion during the Cognitive Computing conference, hosted by the IBM Almaden Institute, 10 May, 2006. Video download available at http://www.almaden.ibm.com/institute/2006/agenda.shtml
[10] Malcolm Ritter, "Brain Electrodes Help Man Speak Again," *The Washington Post*, 1 August 2007.
[11] Kurzweil, *The Singularity is Near*, p. 163.
[12] Richard A.L. Jones, "Rupturing the Nanotech Rapture," *IEEE Spectrum*, vol. 45, no. 6, June 2008, pp. 64-67.
[13] Kurzweil, *The Age of Spiritual Machines*, p. 140.
[14] John Searle, "'I Married a Computer': An Exchange," Searle replies to Kurzweil's response to a review of *The Age of Spiritual Machines: When Computers Exceed Human Intelligence*, by Ray Kurzweil, *The New York Review of Books*, 20 May 1999.
[15] Glenn Zorpette, "Waiting for the Rapture," *IEEE Spectrum*, vol. 45, no. 6, June 2008, pp. 34-35.
[16] Kristen Philipkoski, "Ray Kurzweil's Plan: Never Die," *Wired Magazine* [online], 18 November 2002, available at http://www.wired.com/culture/lifestyle/news/2002/11/56448
[17] Ray Kurzweil and Terry Grossman, *Transcend: Nine Steps to Living Well Forever* (New York: Rodale, 2009), p. xvii.
[18] Kurzweil and Grossman, p. 5.
[19] Gordon Moore, "Tech Luminaries Address Singularity," *IEEE Spectrum Special Report: The Singularity* [online], June 2008, available at http://spectrum.ieee.org/computing/hardware/tech-luminaries-address-singularity
[20] Kurzweil, *The Age of Spiritual Machines*, p. 3.
[21] Steven Pinker, "Tech Luminaries Address Singularity," *IEEE Spectrum Special Report: The Singularity* [online], June 2008, available at http://spectrum.ieee.org/computing/hardware/tech-luminaries-address-singularity

[22] Alfred Nordmann, "Singular Simplicity," *IEEE Spectrum*, vol. 45, no. 6, June 2008, p. 61.
[23] Ken Muneoka, Manjong Han, and David M. Gardiner, "Regrowing Human Limbs," *Scientific American*, vol. 298, no. 4, April 2008, pp. 56-63.
[24] "Genetically Modified "Serial Killer" T Cells Obliterate Tumors in Patients with Chronic Lymphocytic Leukemia, Penn Researchers Report," News Release, 10 August 2011 (Philadelphia: University of Pennsylvania, 2011), available from http://www.uphs.upenn.edu/news/News_Releases/2011/08/t-cells/
[25] Robert Bazell, "New Leukemia Treatment Exceeds 'Wildest Expectations'," *msnbc.com* [online], 10 August 2011, available at http://www.msnbc.msn.com/id/44090512/ns/health-cancer/t/new-leukemia-treatment-exceeds-wildest-expectations/
[26] From CDC fact sheet on diphtheria, available from CDC website at http://www.cdc.gov/vaccines/pubs/pinkbook/downloads/dip.pdf

Chapter 7

[1] Charles Darwin, letter to Hooker, J.D., 9 February 1865, available at http://www.darwinproject.ac.uk/entry-4769
[2] Charles Darwin, *The Life and Letters of Charles Darwin*, vol. 1, edited by Francis Darwin (New York: Basic Books, Inc., 1959), p. 282.
[3] Quoted in Crosbie Smith and M. Norton Wise, *Energy and Empire: A Biographical Study of Lord Kelvin* (Cambridge; New York: Cambridge University Press, 1989), p. 542.
[4] Lawrence M. Krauss and Robert J. Scherrer, "The End of Cosmology?" *Scientific American*, vol. 298, no. 3, March 2008, pp.46-53.
[5] Ray Kurzweil, *The Singularity is Near: When Humans Transcend Biology* (New York: Viking, 2005), p. 361.
[6] Frank Tipler, *The Physics of Immortality: Modern Cosmology, God and the Resurrection of the Dead* (London: MacMillan, 1994), p. 220.
[7] Frank Tipler, "How Significant is an Expanding Universe?" *Closer to Truth* website, available at http://www.closertotruth.com/video-profile/How-Significant-is-an-Expanding-Universe-Frank-Tipler-/1036
[8] Samuel McCracken, "Democratic Capitalism and the Standard of Living 1800-1980," *Modern Capitalism Volume 1: Capitalism and Equality in America*, ed. Peter L. Berger (Lanham, New York and London: Hamilton Press, 1987), p. 18.
[9] Robin E. Bell, "The Unquiet Ice," *Scientific American*, vol. 298, no. 2, February 2008, pp. 60-67.
[10] William Perry, "The Nuclear Chessboard, 2012," a discussion hosted by The Commonwealth Club of California, with George Shultz, William Perry, Sam Nunn and Philip Taubman at the Mark Hopkins Hotel in San Francisco, 23 February 2012.

[11] Photo from Wikimedia Commons, at http://commons.wikimedia.org/wiki/File:Hiroshima_aftermath.jpg
[12] Perry (2012).
[13] Sam Nunn, "The Nuclear Chessboard, 2012," a discussion hosted by The Commonwealth Club of California, with George Shultz, William Perry, Sam Nunn and Philip Taubman at the Mark Hopkins Hotel in San Francisco, 23 February 2012.
[14] Nunn (2012).
[15] Paul W. Ewald, "The Evolution of Virulence," *Scientific American*, vol. 268, no. 4, April 1993, pp. 86-93.
[16] W. Wayt Gibbs, "Synthetic Life," *Scientific American*, vol. 290, no. 5, May 2004, pp. 74-81.
[17] Carl Sagan, *The Demon Haunted World: Science as a Candle in the Dark* (London: Headline, 1996), p. 28.
[18] President Martin Van Buren, quoted by Howard Zinn, *A People's History of the United States* (New York: HarperPerennial, 1980), p. 146.
[19] Frans de Waal, "So Human, So Chimp," *The Human Spark*, written and produced by Graham Chedd, directed by Larry Engel, 57 min., a co-production of Chedd-Angier-Lewis Productions and THIRTEEN in association with WNET.ORG, airdate: 13 January 2010.
[20] Edward O. Wilson, *The Future of Life* (New York: Alfred A. Knopf, 2002), p. 102.
[21] Ken Zweibel, James Mason and Vasilis Fthenakis, "A Solar Grand Plan," *Scientific American*, vol. 298, no. 1, January 2008, pp. 64-73.
[22] Michael Moyer and Carina Storrs, "How Much is Left?" *Scientific American*, vol. 303, no. 3, September 2010, pp. 74-81.
[23] Carl Sagan, *Pale Blue Dot: A Vision of the Human Future in Space* (London: Headline, 1995), p. 53.
[24] Carl Sagan, "The Shores of the Cosmic Ocean," *Cosmos*, directed by Adrian Malone, 1 hr., Cosmos Studios, 1980, DVD Video.
[25] "Chemical Controls," *Scientific American*, vol. 302, no. 4, April, 2010, p. 30.
[26] John Kerry and Teresa Heinz Kerry, *This Moment on Earth: Today's New Environmentalists and Their Vision for the Future* (New York: Perseus, 2007), pp. 42-43, 48.
[27] "Chemical Controls," *Scientific American*, vol. 302, no. 4, April, 2010, p. 30.
[28] Walter C. Willett and Meir J. Stampfer, "Rebuilding the Food Pyramid," *Scientific American*, vol. 288, no. 1, pp. 64-71.
[29] Linda A. Johnson, "KFC, McDonald's Fries and Chicken are Fattier in the U.S. than in Some Other Countries," *The Associated Press*, 12 April 2006.
[30] Steen Stender, Jorn Dyerberg and Arne Astrup, "High Levels of Industrially Produced Trans Fat in Popular Fast Foods," *The New England Journal of Medicine*, vol. 354, 13 April 2006, pp. 1650-1652.
[31] Dariush Mozaffarian et al., "Trans Fatty Acids and Cardiovascular Disease," *The New England Journal of Medicine*, vol. 354, 13 April 2006, pp. 1601-1613.
[32] Anoop Shankar, Jie Xiao and Alan Ducatman, "Perfluoroalkyl Chemicals and

Chronic Kidney Disease in US Adults," *American Journal of Epidemiology*, vol. 174, no. 8, 26 August 2011, pp. 893-900.

[33] Rebecca Renner, "Scotchgard Scotched," *Scientific American*, vol. 284, no. 3, March 2001, p. 18.

[34] H. Wayne Morgan, *Industrial America: The Environment and Social Problems, 1865-1920*, compiled and edited by H. Wayne Morgan (Chicago: Rand McNally College Pub. Co., 1974), p. 71.

[35] Morgan, p. 71.

[36] Kerry and Kerry, p. x.

[37] John Muir, *John Muir: His Life and Letters and Other Writings*, ed. Terry Gifford (London; Baton Wicks; Seattle: Mountaineers, 1996), p. 113.

[38] Photo from Wikimedia Commons, at http://upload.wikimedia.org/wikipedia/commons/f/fb/Muir_and_Roosevelt_restored.jpg

[39] John Muir, *Nature Writings*, ed. William Cronon (New York: The Library of America: distributed to the trade in the U.S. by Penguin Books, USA, 1997), pp. 825-826.

[40] Muir, *Nature Writings*, p. 245.

[41] Muir, *Nature Writings*, p. 825.

[42] Donald K. Grayson, *The Establishment of Human Antiquity* (New York: Academic Press, 1983), p.40.

[43] Rory Carroll, "Pope Says Sorry for Sins of Church," *The Guardian*, 13 March 2000, available at http://www.guardian.co.uk/world/2000/mar/13/catholicism.religion

[44] David Waters, "Sorry Charlie: Church Apologizes to Darwin," *The Washington Post*, 17 September 2008, available at http://onfaith.washingtonpost.com/onfaith/undergod/2008/09/church_of_england_apologizes_t.html

[45] J. Bronowski, *The Ascent of Man* (Boston; Toronto: Little, Brown and Company, 1973), pp. 365-367.

[46] Bronowski, p. 374.

[47] Quoted in Paul R. Ehrlich and Anne H. Ehrlich, *The Population Explosion* (New York: Simon and Schuster, 1990), p. 19.

[48] Ehrlich and Ehrlich, p. 19.

[49] "War," written by Norman Whitfield and Barrett Strong, released 10 June 1970, performed by Edwin Starr.

[50] *Constitution of Ireland* (Dublin: Government Publications Office, 1980), Preamble.

[51] *Constitution of Ireland*, Article 44, Section 1.

[52] Enda Kenny, quoted in: Bob Simon, "The Archbishop of Dublin Challenges the Church," *60 Minutes*, produced by Tom Anderson, 4 March 2012, available at http://www.cbsnews.com/8301-18560_162-57390125/?tag=currentVideoInfo;videoMetaInfo

[53] Enda Kenny, quoted in: "This is a Republic, not the Vatican," *The Irish Times*, 21 July 2011, available at

http://www.irishtimes.com/newspaper/opinion/2011/0721/1224301061733.html
[54] Photo from Wikimedia Commons, at http://upload.wikimedia.org/wikipedia/commons/7/73/God2-Sistine_Chapel.png
[55] Steven Weinberg, "A Designer Universe," *Science and Religion: Are They Compatible?* ed. Paul Kurtz, Barry Karr and Ranjit Sandhu (New York: Prometheus Books, 2003), pp. 39-40.
[56] Muir, *Nature Writings*, p. 826.
[57] "Tentacles of Doom," *Father Ted*, season 2, episode 3, directed by Declan Lowney, 25 min., Hat Trick Productions, 1996, DVD Video.
[58] "Good Luck, Father Ted," *Father Ted*, season 1, episode 1, directed by Declan Lowney, 25 min., Hat Trick Productions, 1995, DVD Video.
[59] Diane Ackerman, *An Alchemy of the Mind: The Marvel and Mystery of the Brain* (New York: Scribner, 2004), p. 54.
[60] Weinberg, p. 39.
[61] Quoted in Hugh Rawson and Margaret Miner, *The Oxford Dictionary of American Quotations,* 2nd Edition (Oxford; New York: Oxford University Press, 2006), p. 606.
[62] Martin Luther, quoted by Stephen Cave, *Immortality: The Quest to Live Forever and How it Drives Civilization* (New York: Crown Publishers, 2012), p.89.
[63] "Good Luck, Father Ted."
[64] Neil deGrasse Tyson, "Holy Wars: An Astrophysicist Ponders the God Question," *Science and Religion: Are They Compatible?* ed. Paul Kurtz, Barry Karr and Ranjit Sandhu (New York: Prometheus Books, 2003), pp. 73-74.
[65] Quoted in Michael Shermer, "The Shamans of Scientism," *Scientific American*, vol. 286, no. 6, June 2002, p. 35.
[66] Stephen Hawking, "Did God Create the Universe?" *Curiosity*, produced by Alan Eyres et al., 45 min., a production of Darlow Smithson Productions for Discovery Channel, airdate: 7 August 2011.
[67] John Muir, quoted by Linnie Marsh Wolfe, *Son of the Wilderness: The Life of John Muir* (Madison, Wisconsin: University of Wisconsin Press, 2003), p. 95.
[68] Carl Sagan, *Cosmos* (London: Macdonald Futura Publishers, 1981), p. 29.
[69] Carl Sagan, *Billions and Billions: Thoughts on Life and Death at the Brink of the Millennium* (New York: Random House, 1997), pp. 214-215.
[70] Sagan, *Billions and Billions*, p. 215.
[71] Robert Ingersoll, quoted by Michael Shermer, *Why People Believe Weird Things: Pseudoscience, Superstition, and Other Confusions of Our Time* (New York: W.H. Freeman and Company, 1997), p.82.
[72] Stephen Mitchell, translator, *Gilgamesh: A New English Version* (New York, London, Toronto, Sydney: FREE PRESS, 2004), p. 50.
[73] T.S. Eliot, "Little Gidding," *Collected Poems: 1909-1962* (San Diego; New York; London: Harcourt Brace Jovanovich, 1963), p. 208.

Chapter 8

[1] Carl Sagan, "The Shores of the Cosmic Ocean," *Cosmos*, directed by Adrian Malone, 1 hr., Cosmos Studios, 1980, DVD Video.
[2] *Dark Star*, directed by John Carpenter, 1 hr. 23 min., Jack H. Harris Enterprises; University of Southern California (USC), 1974, DVD Video.
[3] Carl Sagan, *Pale Blue Dot: A Vision of the Human Future in Space* (London: Headline, 1995), pp. 371-377.
[4] Carl Sagan, *Cosmos* (London: Macdonald Futura Publishers, 1981), p. 5.
[5] Carl Sagan, "Let's Go to Mars Together," *The Planetary Report*, vol. 6, no. 4, July/August 1986, pp. 8-9.
[6] Neil deGrasse Tyson, "Can We Live Forever?" *NOVA ScienceNOW*, written and produced by Vincent Liota, 55 min., produced for WGBH/Boston by NOVA, airdate: 26 January 2011.
[7] Photo by NASA, retrieved from Wikimedia Commons at http://commons.wikimedia.org/wiki/File:Nichelle_Nichols,_NASA_Recruiter_-_GPN-2004-00017.jpg
[8] Jesper L. Anderson, Peter Schjerling and Bengt Saltin, "Muscle and the Elderly," *Scientific American*, vol. 283, no. 3, September 2000, p. 54.
[9] Kathryn Brown, "A Radical Proposal," *Scientific American Presents*, vol. 11, no. 2, Summer 2000, p. 39.
[10] S. Jay Olshansky and Bruce A. Carnes, *The Quest for Immortality: Science at the Frontiers of Aging* (New York; London: W.W. Norton & Company: 2001), p.128.
[11] Ralf Dahm, "Dying to See," *Scientific American*, vol.291, no. 4, October 2004, p. 82-89.
[12] Lisa Melton, "AGE Breakers," *Scientific American*, vol. 283, no. 1, July 2000, p. 16.
[13] Quoted by Karen Hopkin, "Making Methuselah," *Scientific American Presents: Your Bionic Future*, vol.10, no. 3, 1999, pp. 32-37.
[14] Vojo Deretic and Daniel J. Klionsky, "How Cells Clean House," *Scientific American*, vol. 298, no. 5, May 2008, pp. 74-81.
[15] Pramod K. Srivastava, "New Jobs for Ancient Chaperones," *Scientific American*, vol. 299, no. 1, July 2008, pp. 50-55.
[16] Roy Walford, "Fat and Happy," *Scientific American Frontiers*, written, produced and directed by John Angier, David Huntley and Andy Liebman, 55 min., The Chedd-Angier Production Company, Inc., airdate: 1 May 2001.
[17] Alan Alda, "Fat and Happy."
[18] Bradley Willcox et al., "FOXO3A Genotype is Strongly Associated with Human Longevity," *PNAS*, vol. 105, no. 37, 16 September 2008, pp. 13987-13992.
[19] Barbara Juncosa, "Is 100 the New 80?: Centenarians Studied to Find the Secret of Longevity," *Scientific American* [online], 28 October 2008, available at http://www.scientificamerican.com/article.cfm?id=centarians-studied-to-find-

the-secret-of-longevity
[20] Gerald J. Gruman, *A History of Ideas about the Prolongation of Life* (New York: Springer Publishing Company, 2003), p. 6.
[21] Gruman, p. 18.
[22] Roberta Sarfatt Borkat, "Pride, Progress, and Swift's Struldbruggs," *Durham University Journal*, vol. 68, June 1976, pp. 126-134.
[23] Brian Williams, *NBC Nightly News*, airdate: 6 February 2012.
[24] Jason Leigh, "Can We Live Forever?" *NOVA ScienceNOW*
[25] Quoted by Sebastian Seung, *Connectome: How the Brain's Wiring Makes Us Who We Are* (Boston; New York: Houghton Mifflin Harcourt, 2012), p. 252.
[26] Christof Koch, "Consciousness," a lecture presentation delivered by Christof Koch during the Cognitive Computing conference, hosted by the IBM Almaden Institute, 11 May, 2006. Video download available at http://www.almaden.ibm.com/institute/2006/agenda.shtml
[27] James E. Zull, *From Brain to Mind: Using Neuroscience to Guide Change in Education* (Sterling, Virginia: Stylus Publishing, LLC., 2011), p. 89.
[28] "NIH Launches the Human Connectome Project to Unravel the Brain's Connections," *NIH News*, 15 July 2009, available at http://www.nih.gov/news/health/jul2009/ninds-15.htm
[29] Matthew Knight, "Mapping Out a New Era in Brain Research," *CNN* [online], 9 April 2012, available at http://www.cnn.com/2012/03/01/tech/innovation/brain-map-connectome/index.html
[30] Steven Rose, *The Future of the Brain: The Promise and Perils of Tomorrow's Neuroscience*, (Oxford; New York: Oxford University Press, 2005), pp. 216-220.
[31] Sandra Blakeslee, "Monkey's Thoughts Propel Robot, A Step That May Help Humans," *New York Times* [online], 15 January 2008, available at http://www.nytimes.com/2008/01/15/science/15robo.html?_r=1&pagewanted=all
[32] Matt McGrath, "Monkeys' Brain Waves Offer Quadriplegics Hope," *BBC News* [online], 6 October 2011, available at http://www.bbc.co.uk/news/science-environment-15197124
[33] Ray Kurzweil, *The Age of Spiritual Machines: When Computers Exceed Human Intelligence* (New York: Penguin Books, 1999), pp. 146-150, 194, 205-206.
[34] Alan Johnston, "Italy Quake Homeless in Emergency Shelters," *BBC News* [online], 21 May 2012, available at http://www.bbc.co.uk/news/world-europe-18140543
[35] Mark Dezzani, "New Earthquake in Northern Italy Kills 16," *BBC News* [online], 29 May 2012, available at http://www.bbc.co.uk/news/world-europe-18247659
[36] Donald Rumsfeld, "DoD News Briefing – Secretary Rumsfeld and Gen. Myers," News Transcript, 12 Feb 2002 (Washington, D.C.: Federal News Service, Inc., 2002), available from

http://www.defense.gov/transcripts/transcript.aspx?transcriptid=2636

[37] Stephen Cave, *Immortality: The Quest to Live Forever and How it Drives Civilization* (New York: Crown Publishers, 2012), pp. 131-134.

[38] John Searle, "What is the Nature of Personal Identity?" *Closer to Truth* website, available at http://www.closertotruth.com/video-profile/What-is-the-Nature-of-Personal-Identity-John-Searle-/439

[39] Quoted by Stephen Mitchell, translator, *Gilgamesh: A New English Version* (New York, London, Toronto, Sydney: FREE PRESS, 2004), pp. 159.

[40] Quoted by Mitchell, pp. 198-199.

[41] T.S. Eliot, "Little Gidding," *Collected Poems: 1909-1962* (San Diego; New York; London: Harcourt Brace Jovanovich, 1963), p. 208.

[42] Cave, p. 250.

[43] Cave, p. 286.

[44] Paul Davies, "That Mysterious Flow," *Scientific American*, vol. 287, no. 3, September 2002, p. 47.

[45] Davies, p. 47.

[46] John R. Searle, *The Mystery of Consciousness* (New York: The New York Review of Books, 1997), p. xiv.

[47] Jesse Bering, "The End?: Why So Many of Us Think Our Minds Continue On After We Die," *Scientific American Mind,* October/November 2008, p. 37.

[48] Bering, p. 41.

[49] Christof Koch, *The Quest for Consciousness: A Neurobiological Approach* (Englewood, Colorado: Roberts and Company Publishers, 2004), p. 299.

[50] Benjamin Franklin, quoted in Hugh Rawson and Margaret Miner, *The Oxford Dictionary of American Quotations,* 2nd Edition (Oxford; New York: Oxford University Press, 2006), p. 165.

[51] Seung, p. 236.

[52] Elizabeth Kübler-Ross, M.D., *On Death and Dying* (New York: Macmillan Publishing Company, 1970), p. 16.

[53] John Searle, "Beyond Dualism," a lecture presentation delivered by John Searle during the Cognitive Computing conference, hosted by the IBM Almaden Institute, 11 May, 2006. Video download available at http://www.almaden.ibm.com/institute/2006/agenda.shtml

[54] Rudyard Kipling, "The Last Department," *Departmental Ditties and Ballads and Barrack Room Ballads* (New York: Doubleday, Page & Company, 1915), p. 49.

Printed in Great Britain
by Amazon